Invitation to Organic Chemistry

GRIGORIY A. SEREDA

LINUS Learning

Published by Linus Learning

Ronkonkoma, NY 11779

ISBN 10: 1-60797-722-2

ISBN 13: 978-1-60797-722-3

Printed in the United States of America.

This book is printed on acid-free paper.

Print Number 5 4 3 2 1

TABLE OF CONTENTS

TO THE READER

This book is intended as a supplementary material for students starting to learn organic chemistry. Regardless of their educational goals, students often struggle to navigate through hefty textbooks, scary formulas, and baseless terrifying rumors perpetuated by students who have failed organic chemistry in the past. The presented material will help fit organic chemistry into the reader's prior knowledge, world view, and flesh out connections to other chemistry disciplines, biology, and everyday life. The subject of organic chemistry will be related to many career options available to the learner after completion of this course. This book will not only offer a roadmap for learning organic chemistry depending on the student's career goals, but help in choosing, shaping, and achieving those goals. Each topic is presented with a reference to the common pitfalls preventing students from making rapid progress in the organic chemistry class.

In each chapter of the book, after discussing a specific topic of organic chemistry, we will briefly revisit the material from the following perspectives:

a) *What can go wrong?* Here we will discuss specific learning challenges commonly faced by students in a classroom, based on the author's 15+ year experience teaching organic chemistry at all levels, ranging from middle school to Ph.D. graduate students.

b) *Pushing the limits.* The reader will be invited to critically think about limitations of the rules and concepts learned in the chapter. Why, when, and how well does the rule work? When and why does it not? How to take a practical advantage of it? This section will show the reader further perspectives of building knowledge of organic chemistry.

c) *Practice problems.* These problems will be supplemented by a CD with computer animations of most common organic reaction mechanisms developed by the author.

d) *Making connections.* This section will list topics from previous chapters necessary for understanding the current material. We will also indicate which of the following chapters are based on the material covered here.

Briefly, this book will friendly invite the reader to the world of organic chemistry, show its role in our everyday life, where it stands among other scientific disciplines, and help the reader decide where and how to go from there depending on the personal interest and career goals.

Chapter 1

Language of Organic Chemistry

First, we need to appreciate the inherent and amazing uniqueness of atoms and any atomic structure made up by positively charged nuclei and negatively charged electrons. Despite the significant electrostatic attraction between nuclei and electrons, atoms and molecules do not collapse, ensuring stability of the Universe in its current form. According to quantum physics, at the atomic size scale, particles behave like waves, so electrons do not fall on the nuclei, because *waves cannot fall.*

According to a set of approximations justified in appropriate courses of quantum chemistry, energy and location of each electron is described by a wave of a certain shape and location. Mathematical expression of such wave is called *orbital*, which describes up to two electrons with opposite magnetic moments (spins). Sometimes, for more advance descriptions outside of this course, a unique orbital is assigned to each electron (spin-orbital). Amplitude (absolute value) of the orbital at each point of space shows the probability (strictly speaking, density of probability) of finding an electron at that point. Notably, at some locations orbitals are positive, while at other locations they are negative, which means that interacting orbitals may either amplify, or cancel each other depending on the specific location. The chemical implication is that bringing together several atoms or molecules leads to redistribution of electrons between nuclei. Some orbitals enhance each other between atomic nuclei and are called *bonding orbitals*. Electrons described by such orbitals are shifted toward locations between atomic nuclei and are pulling them together like "electrostatic glue", therefore, strengthening a chemical bond. Other orbitals cancel (or weaken) each other between atomic nuclei and are called *antibonding orbitals*. Electrons described by such orbitals are shifted toward locations behind atomic nuclei and are pulling them apart, therefore, weakening a chemical bond. Some electrons, however, do not significantly redistribute between the nuclei of interacting atomic structures and remain "*unshared*" electrons. A pair of unshared electrons is described by a *non-bonding orbital*, and is usually called a "*lone electron pair*". The term "occupied orbital" means "there are two electrons described by this orbital". The term "unoccupied orbital" means "there are currently no electrons described by this orbital". This orbital defines a space where an electron can be placed, which is useful to describe interaction of the particle with electron donors. The term "singly occupied orbital" means "there is one electron described by this orbital". We need to keep in mind that unlike electrons, orbitals are not real objects. They are mathematical functions used to describe real electrons.

To summarize this, each electron pair described by a bonding orbital, corresponds to one chemical bond of a certain type (*covalent bond*) if all antibonding orbitals are unoccupied. This concepts lets us introduce very powerful tools to visualize distribution of electrons in a molecule *(Lewis structure)* and their redistribution during chemical reactions *(Curved arrow notation)*. **These two tools truly consist the language of organic chemistry.**

Lewis structure is a conventional chemical structure (bonds are shown as lines between connected atoms), where electrons on bonding and antibonding orbitals are shown as dots. Since an *electron pair on a bonding orbital* corresponds to a covalent bond, it is always shown *near to the bond line,* which stands for that bond. Since an *electron pair on a non-bonding orbital* is not shared between atoms, it is always shown *near to a specific atom.* For simplicity, the dots on chemical bonds are usually omitted unless they are involved in a reaction mechanism that has to be shown. If lone electrons pairs are also omitted, the Lewis structure becomes a structural formula. However, if we show a lone electron pair on an atom, we must show all of them.

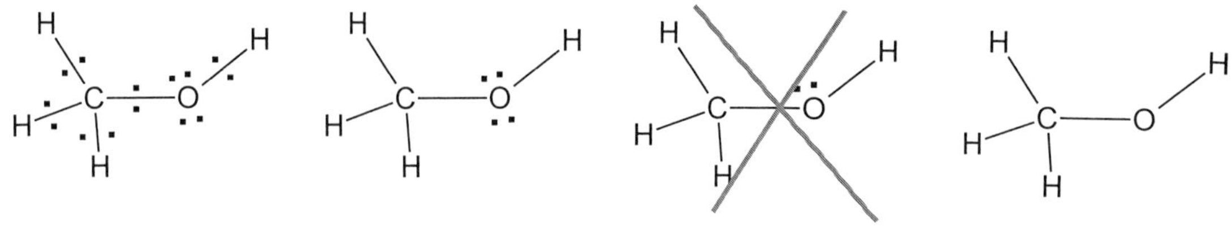

Lewis structures and the structural formula of methanol

However, while using Lewis structures, we need to realize that an actual bonding orbital is never localized (or even mostly localized) between a specific pair of atoms except for diatomic molecules and isolated multiple bonds, so chemical bonds should not be considered as isolated entities. Now, with the knowledge of approximations and assumptions of the concept "Lewis structure", we are ready to understand both its value and limitations.

Lewis structures should be used for:

a) Description of connectivity of atoms in a molecule (the sequence in which atoms are connected).

b) "Bookkeeping" of electrons and charges in the molecule.

c) Prediction of the shape of the molecule (molecular geometry).

d) Qualitative estimation of the overall strength of bonding in the molecule (potential energy), and its relationship with the molecular geometry.

e) Identification of reactive centers in the molecule.

f) Qualitative estimation of chemical reactivity of a molecule (with caution). The art of such estimation in order to design reactions of practical use is mostly what organic chemistry is about.

Lewis structures should not be directly used for:

a) Description of any part of molecule with a significant contribution of occupied antibonding orbitals (e.g. O_2, HHe), which is also often true for organic molecules in their excited states.

b) Phenomena based on energies of individual orbitals (potential of ionization, electronic spectra, etc.).

c) Structures, where placing two electrons on each covalent bond would violate the doublet and octet rules (e.g. B_2H_6, ClI_5^+).

d) Parts of a molecule, where occupied and unoccupied orbitals mostly localized on atoms (lone electron pairs, atoms of the 2nd period with incomplete electron octet, atoms of 3rd+ periods with unoccupied d- or f-orbitals) or pairs of atoms (double and triple bonds), significantly overlap (e.g. CH_2=CH-CH_2^+, CH_2=CH-CH=CH_2, CH_3-C(O)NH_2, etc.).

Later we will learn that limitation d), and less efficiently – limitations a) and c), can be remediated using the concept of *resonance structures*. Now, we will look at the first application of Lewis structures as a tool for calculation of *formal charges* – a necessary element of a correct structural formula.

Formal charge is the actual charge of an atom that would have been there had all covalent bonds been cleaved and the atom had kept one electron from each bond.

Formal Charge is calculated as the difference between the nuclear charge, corresponding to the valence electrons and the number of electrons, contributing to the atomic charge. The number of electrons left contributing to the atomic charge on the atom can be easily calculated by counting two electrons from each lone electron pair (LP, both electrons belong to this atom) and one electron from each covalent bond (CB, one out of two shared electrons belongs to this atom). Since redistribution of non-valence (core) electrons is negligible versus their valence counterparts, we consider only valence electrons in Lewis structures, and count only them for calculation of the formal charge (Ch). Therefore, we need to count only the part of the nuclear charge (Nucl) that corresponds to the valence electrons, which equals the nucleus's group number in the periodic table. Here is a simple formula to help you with these calculations: Ch = Nucl – CB – 2*LP

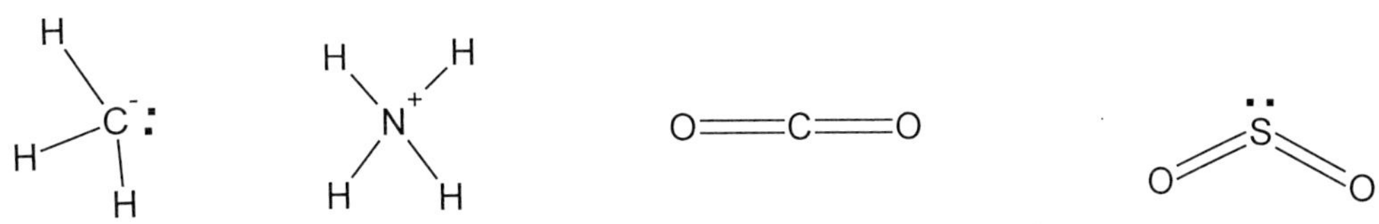

Carbon:	Nitrogen:	Carbon:	Sulfur:
Nucl = 4	Nucl = 5	Nucl = 4	Nucl = 6
CB = 3	CB = 4	CB = 4	CB = 4
LP = 1	LP = 0	LP = 0	LP = 1
Ch = -1	**Ch = +1**	**Ch = 0**	**Ch = 0**

Examples for calculating formal charges and the number of lone electron pairs

In order to take full advantage of Lewis structures and make sure that your drawing is accepted by most professors and grading programs as correct answers, please, follow the following guidelines:

1. Always show the formal charge. It is essential for any structural formula, not only Lewis structures.

2. Always show the electrons essential for the idea you are trying to get across, using the Lewis structure. It is a part of the language of organic chemistry after all.

3. Don't show non-valence electrons. Think of the fun you may have placing 92 dots around the atom of uranium in the formula UF_6.

4. Show unshared electrons, if it's essential for what you are writing the structure for.

5. If you have shown any of shared or unshared electrons, you must show all unshared electrons. Placing just one instead of two lone electron pairs around oxygen in the structure of H_2O may mislead you regarding the molecular geometry.

6. The doublet for the elements of the 1st period and the octet for the elements of the 2nd period may not be exceeded, but do not necessarily have to be complete. A five- or six-valent carbon in your homework or test may give your professor a strong urge to assign you "F".

Molecular geometry. The next powerful application of Lewis structures is predicting molecular shapes based on a very straightforward premise, that electrostatic repulsion keeps alike-charged electrons (and, therefore, electron-rich areas such as covalent bonds and lone electron pairs) as far from each other as possible. While involvement of d- and f orbitals for transition elements and lanthanides can lead to more, than one answer, for predicting molecular shapes of most organic molecules, that is all you need to know. There are three types of electronic environment to consider:

a) Two electron-rich areas end up on the opposite sides of the atom. If both those areas are covalent bonds, the bond angle is about 180°. *This is linear geometry.*

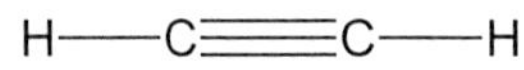

Acetylene

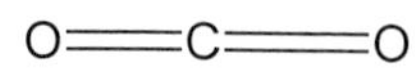

Carbon dioxide

Examples of molecules with linear geometry around carbon

b) Three electron-rich areas end up at the vertexes of a planar triangle. If all those areas are covalent bonds, the atom has a planar trigonal environment with the bond angle about 120°. *This is trigonal planar geometry.*

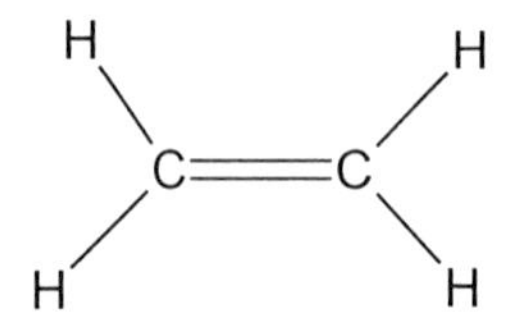

Ethylene

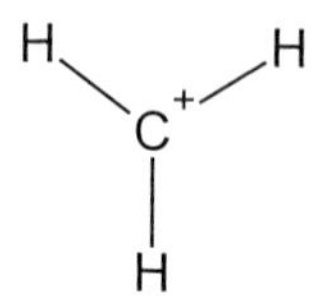

Methyl cation

Examples of molecules with trigonal planar geometry around carbon

If one of those areas is a lone electron pair, the remaining two bonds are pushed by that electron pair out the linear arrangement to an angle slightly less, than 120o. This angle is less than 120°, because the lone electron pair has higher energy, than a bonding pair, stays further from the nucleus, takes more space, and pushes on the two remaining bonding pairs harder. This is bent geometry.

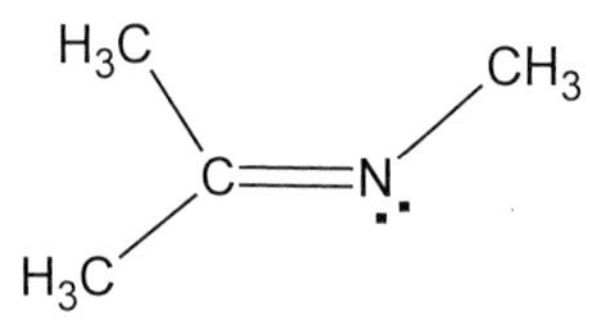

Acetone methylimine (geometry at nitrogen)

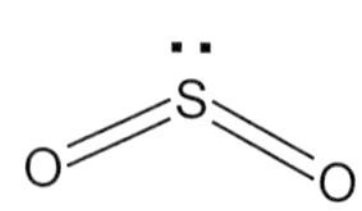

Sulfur dioxide

Examples of molecules with bent geometry

c) Repulsion of four electron-rich areas pushes them to the vertexes of a tetrahedron, which corresponds to the bond angle of about 108.5°. *This is tetrahedral geometry.*

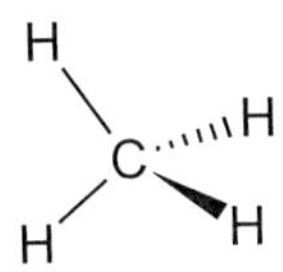

Methane

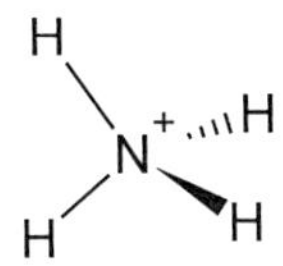

Ammonium

Examples of molecules with tetrahedral geometry

Note: In both structures, bonds located in the plane of the drawing are shown by solid lines. Bonds pointing at us are shown by solid wedges. Bonds pointing away from us are shown by dashed wedges. This is a common way of showing molecules in three dimensions.

If one of those areas is a lone electron pair, the remaining three bonds are pushed by that electron pair out of the planar arrangement to an angle slightly less, than 108.5°. This is trigonal pyramid geometry.

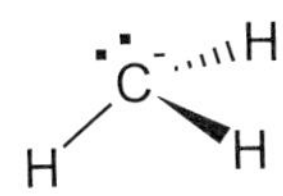

Methyl anion

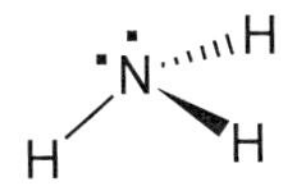

Ammonia

Examples of molecules with trigonal pyramidal geometry

If two of those areas are lone electron pairs, the remaining two bonds will form an angle slightly less, than 108.5°. This is again bent geometry.

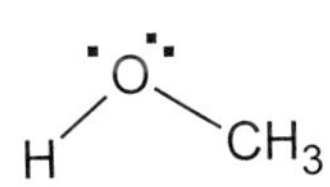

Methanol (geometry at oxygen)

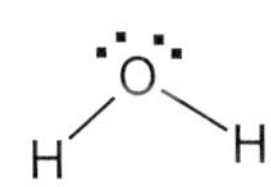

Water

More examples of molecules with bent geometry

Hybridization. The electrons necessary for chemical bonding are supplied to the bonding orbitals by the bonded atoms. In isolated atoms, just like in molecules, these electrons are described by orbitals of different types. Relative contributions of those orbitals to the bond formation significantly influences its properties. For elements of the 2nd period (e.g. C,N,O), one s- (lower in energy sphere shaped) and three p- (higher in energy dumbbell shaped) orbitals make by far major contribution to the chemical bonds of that atom. The spherical atomic s-orbital always contributes to the bonding orbitals significantly located between this atom's nucleus and the nuclei of the atoms it is bound to. Rotation of two parts of the molecule around the line connecting the nuclei does not break such bond, which is called a *σ-bond*. Bonding and antibonding orbitals are calculated from interacting atomic orbitals. The relative contributions (usually, coefficients of a linear combination) of s- and p-atomic orbitals to the calculation of bonding orbitals (describing σ-bonds) and non-bonding orbitals (describing lone electron pairs) around an atom, determine hybridization type of that atom, if those contributions are not very different for all σ-bonds and lone electron pairs.

The total number of atomic orbitals contributed by an atom to form bonds with that atom usually equals the number of those bonds. The reason is that mathematical formalism behind calculation of bonding and antibonding orbitals from atomic orbitals dictates that the total number of new orbitals equals the number of initial atomic orbitals. The total number of bonding and antibonding orbitals is twice the number of bonds (one bonding and one antibonding per bond). This number matches the number of atomic orbitals contributed by one orbital from each bonded atom per bond (twice the number of bonds). Also, the atom contributes one atomic orbital per each non-bonding orbital, describing a lone electron pair.

How many of s-and p-orbitals from each atom are most convenient to consider for each chemical bond depends on the geometry around that atom and determines its hybridization type. It should be understood that *hybridization is not a real phenomenon or process, but rather a theoretical treatment of a certain geometrical configuration around an atom.*

a) Two electron-rich areas (σ-bonds) on the opposite sides of the atom. Two atomic orbitals of the atom (one s- and one p-) contribute to the exactly description of two σ-bonds, making the linear geometry characteristic for the *sp-hybridization*. That atom is called "*sp-hybridized*". Examples: carbon in acetylene, carbon dioxide.

b) Three electron-rich areas (σ-bonds) at the vertexes of a planar triangle. Three atomic orbitals of the atom (one s- and two p-) contribute to the description of three σ-bonds, making the trigonal planar geometry characteristic for the *sp^2-hybridization*. If one of those areas is a lone electron pair, the sp^2 hybridization corresponds to the bent geometry. Examples: the atom in ethylene, methyl cation, sulfur in sulfur dioxide, nitrogen in acetone methylimine.

c) Four electron-rich areas at the vertexes of a tetrahedron. Four atomic orbitals of the atom (one s- and three p-) contribute to the description of four σ-bonds, making the tetrahedral geometry characteristic for the *sp^3-hybridization*. If one of those areas is a lone electron pair, the sp^3-hybridization corresponds to the trigonal pyramid geometry. If two of those areas are lone electron pairs, the sp^3-hybridization corresponds to the bent geometry. Examples: carbon in methane, nitrogen in ammonium, oxygen in methanol, oxygen in water.

The concept of Lewis structures is based on the assumption that electron pairs are mostly localized at one or two atoms (*two centered bond*), which never holds perfectly true in real molecules. Fortunately, this assumption usually works well enough to let us draw meaningful Lewis structures and estimate basic molecular properties. The more the real molecule deviates from the Lewis structure assumption, the more additional effects we need to consider for an adequate description of the molecule.

Resonance structures. Every so often, electrons are so significantly shifted from a two centered bond, that one Lewis structure does not provide an adequate description of a molecule or ion. Fortunately, in most cases, the electrons are shifted toward other chemical bonds in a predictable fashion, so another Lewis structure can be drawn for the same molecule. A set of Lewis structures assigned to the same molecule or ion is called a set of *resonance structures*. While neither single structure from the set provides an adequate description of a molecule, the "truth" lies in between the descriptions given by the contributing resonance structures. For instance, the left resonance structure of the allyl cation shows positive charge on the left carbon and a carbon-carbon double bond on the right. The second resonance structure shows positive charge on the right carbon and a carbon-carbon double bond on the left. We will discuss later role of the curved arrows on the picture.

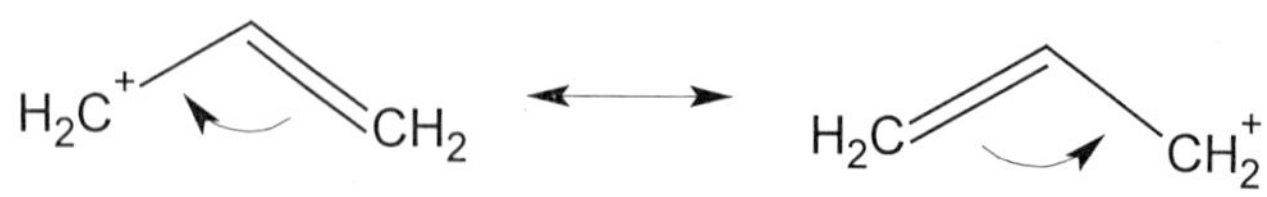

Resonance structures of the allyl cation

This set of resonance structure should be read: "The positive charge is equally distributed between the left and right carbons (but not the middle one), and both carbon-carbon bonds are between a single and a double bond (bond order equals ½)".

In the above example, both resonance structures provide equal contributions to the description of charges and bond orders. However, in many cases some structures are more "important", than others. Thus, a non-ionic structure of formamide (on the left of the scheme below) has a higher contribution (weight) to the description of the molecule, than the ionic one (on the right).

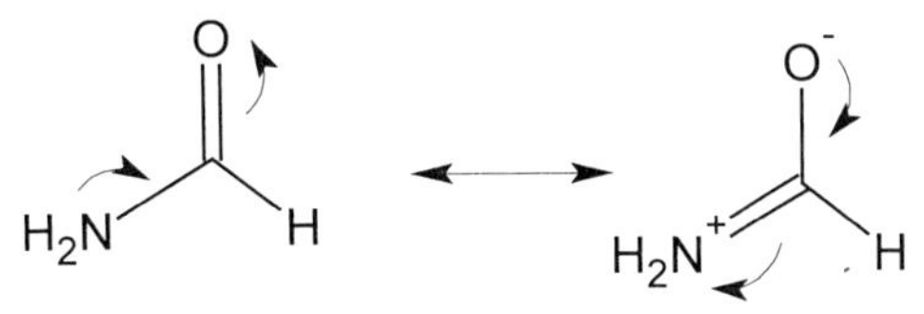

Resonance structures of formamide

The qualitative assessment of the relative contributions of resonance structures requires some knowledge of functional groups in organic chemistry.

Structural elements and functional groups. Understanding the concept of functional groups is, probably, the most powerful tool for avoiding unnecessary memorization. While definitions of the functional group vary from one source to another, you should consider *functional group just as a familiar part of a molecule to look at first*. Chemical properties of a molecule are defined by properties of structural elements, interaction of structural elements, and, in a small extent – by the molecule as a whole. For the purpose of nomenclature, some structural elements are called functional groups, some are called non-functional groups, some – are not considered as "groups". Regardless of terminology, *identifying familiar structural elements is the very first thing to do* when you look at a chemical structure. By the presence of a specific functional group, organic compounds are divided into several classes, chemistry of which will be discussed in the following chapters. The list of most common functional groups and the corresponding rules of nomenclature are usually well presented in any textbook of organic chemistry. *The next thing to consider is interaction of each structural element with other parts of the molecule,* starting from adjacent ones. For instance, in the following molecule of a polyunsaturated fat, we can identify an ester group (–O-C=O), hydrocarbon chains, and C=C bonds.

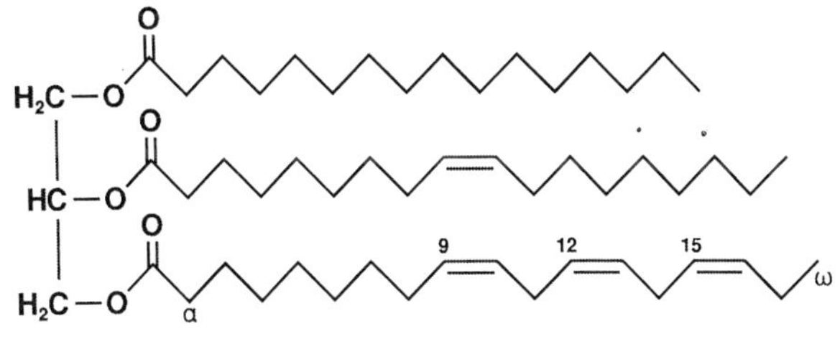

Structure of an unsaturated fat

Therefore, we should expect from this compound at least three types of chemical reactions plus possible chemical properties arising from interaction between those three structural elements. We will revisit this structure in the following chapters and will be able to specifically discuss its chemical reactions. Assignment of a compound to specific classes allows to narrow down chemical properties brought about by each structural element. According to the definitions of the classes of organic compounds, we should place the molecule above to the class of esters (as opposed to aldehydes or ketones), which narrows down chemistry of the carbonyl group (C=O). At the same time, the compound should also be placed to the class of olefins (compound containing C=C bonds). Finally, presence of C=C bonds in specific positions with respect to one another brings about chemical properties not characteristic for simple olefins. Chemical behavior of each carbon of the hydrocarbon chain also depends on the proximity of the carbonyl group and the C=C bond.

Isomerism. Isomerism is ability of the same set of atoms to form a molecule in more than one way. Isomers always have same molecular formula and same elemental composition (percentage of each element). The simplest type of isomers are structural (constitutional) isomers, in which atoms are connected in a different order. The list of atoms, their connected neighbors, and multiplicity (single, double, or triple) of connecting bonds constitutes the connectivity matrix, with is always different for different structural isomers. To be able to distinguish structural isomers, you do not need to write connectivity matrices. Identifying at least one difference in the order of connecting atoms is sufficient to term the compounds structural isomers. For instance, the structure C_4H_{10} on the left (butane) does not contain any carbons connected to three carbons, while another C_4H_{10} structure on the right (2-methylpropane) contains such carbon. Therefore, those two structures are structural isomers. Another example of different connectivity in this pair: The molecule of butane has two carbons connected to three hydrogens, while 2-methylpropane has three such atoms.

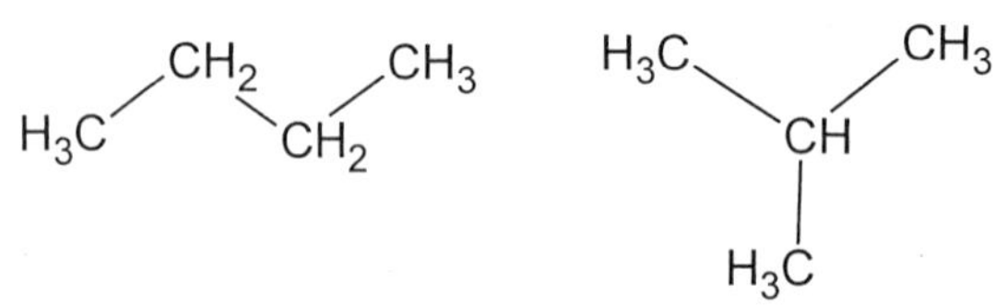

Butane **2-Methylpropane**

Structural isomers of C_4H_{10}

Curved arrow notation. The first question that may come into mind: "Why are those arrows *curved?*" The curved arrow stands for only one very specific thing: *displacement of one pair of electrons,* and is curved to avoid confusion with other types of arrows used for many different purposes (direction of a chemical reaction, direction of a dipole, a coordination bond, pointer to a specific element of the drawing, etc.). Since one electron pair is responsible for a chemical bond, and electrons are negatively charged, displacement of electron pairs always results in *redistribution of bonds and charges.* The curved arrows are used for two different purposes that should not be confused;

1. Redistribution of bonds and formal charges in the process of a single reaction step (a real process).

2. Redistribution of bonds and formal charges while relating one resonance structure to another (not a real process).

Once the starting structure and curved arrows are shown, you need to follow four steps to figure out the related resonance structure, or structure of the product of the next reaction step:

1. If the space between two atoms loses electrons, one covalent bond breaks.

2. If the space between two atoms gains electrons, one covalent bond forms.

3. If any atom loses electrons, its charge increases by 1.

4. If any atoms gains electrons, its charge decreases by 1.

In the following example of the curved arrow notation, the bonds R_1-C and N-O break, because they lose two electrons. The R_1-N bond forms, because the space between these groups gains two electrons. The carbon connected to nitrogen, loses a shared pair of electrons, of those only one electron belonged to that carbon, increases its charge by 1, and becomes positively charged. Oxygen loses a shared pair of electrons, of those only one electron belonged to oxygen, and gains a lone electron pair. Overall, that oxygen gains one electron, decreases its charge by 1, and becomes neutral. Nitrogen gains a shared pair of electrons, loses a shared pair of electron, and, therefore stays neutral. R_2 and both hydrogens are not affected by the curved arrows. Therefore, they keep their charges and bonding.

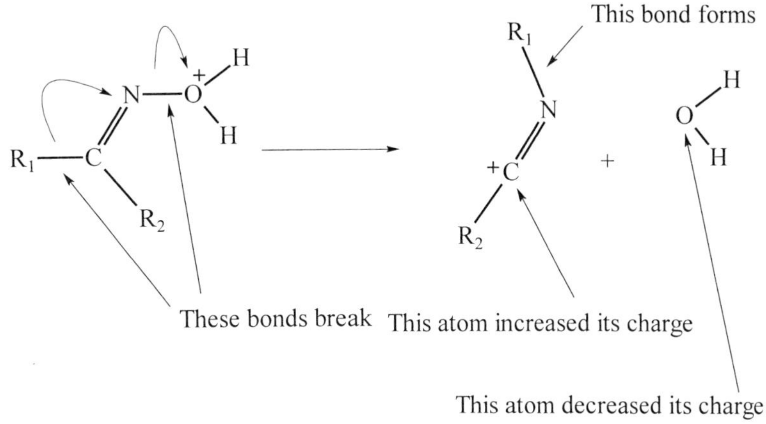

An organic reaction mechanism is shown using the curved arrow notation

The curved arrow notation is used for two different purposes: 1. Actual chemical reaction; and 2. Relationship between resonance structures. The first application is shown in the reaction mechanism above, and the second one was used for the resonance structures of the allyl cation and formamide.

Fish hook notation. "Why are those arrows curved?" The fish hook stands for only one very specific thing: displacement of one electron.

In the following example of the fish hook notation, the Br-Br bond breaks, because it loses both electrons. The C-Br bond forms, because the space between these groups gains two electrons. The carbon loses one electron and gains a shared pair of electrons, of those only one belongs to carbon. Overall, this carbon gains or loses no electrons, and stays neutral. Same is true about the Br that is forming a bond with carbon. The second Br gains one electron and loses a shared pair of electrons, of those only one belongs to Br. Overall, this Br gains or loses no electrons, and stays neutral. All other carbons and hydrogens are not affected by the fish hooks. Therefore, they keep their charges and bonding.

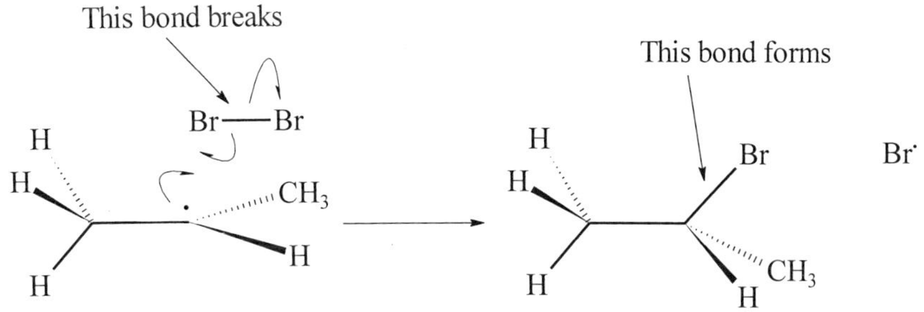

An organic reaction mechanism is shown using the fish hook notation

The web-based animations of curved arrow notations "(http://ochem.orgfree.com/)" lets you practice to depict many other reaction mechanisms by curved arrows or fish hook arrows.

Acids and bases. Perhaps, the simplest application of the curved arrow notation is representing donation of an electron pair from a Lewis base to a Lewis acid, and donation of a proton from a Bronsted acid to a Bronsted base as shown below.

Conjugate base is a deprotonated acid. Conjugate acid is a protonated base.

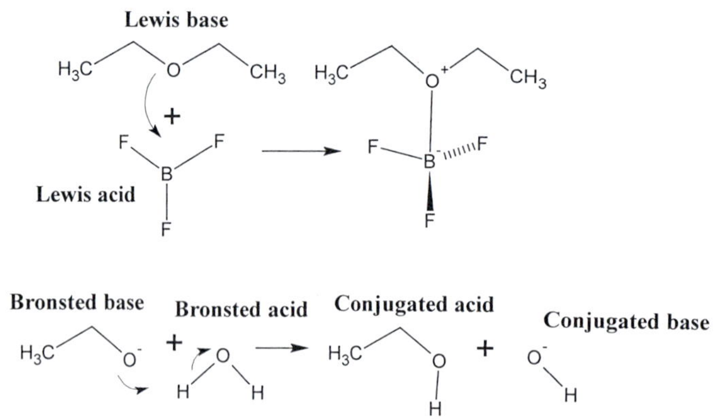

Examples of acid-base reactions

These acid-base interactions are very common steps of organic reaction mechanisms, so their understanding helps to avoid memorization of those steps over and over. For instance, the first reaction above precedes electrophilic substitution, and the second one activates an alcohol for substitution or elimination.

The second important application of the acid-base theory in organic chemistry is to estimate position of the acid-base equilibrium and, therefore, properly generate the reactive particles of interest. First, we need to remember, that an acid –base equilibrium is shifted from a stronger acid (with lower pK_a and weaker conjugate base) to a weaker acid (with higher pK_a and stronger conjugate base). Simply by looking at the pK_a tables, we can tell that OH^- (pK_a=15.7 for H_2O) can only partially deprotonate most alcohols (pKa~15, comparable acidity with H_2O), but almost completely deprotonates carboxylic acids (pKa ~5, much stronger acids, than H_2O).

In an aqueous solution, acidity or basicity mostly translates to changes in the pH value ranging from -1 to 15. The second important take-home lesson from general chemistry is that concentration of an acid becomes equal to the concentration of its conjugate base when the pH value is made equal to the pK_a of that acid. For instance, in an aqueous solution, we can deprotonate alcohols (pK_a~15) at most by half if we max out the pH value, we can significantly deprotonate carboxylic acids (pK_a~5) by raising pH above 5, and we cannot noticeably deprotonate terminal alkynes (pK_a~25). In order to do that, we would have to raise pH above 25, with is impossible.

What can go wrong?

1. Application of chemistry of the reaction center to a structure containing this center. It is important to identify the functional group of interest no matter how it is presented and oriented in a structure that also can be drawn in a variety of ways. To practice, try to write the acid-base equilibrium between the carboxylic group and OH^- for each structure below.

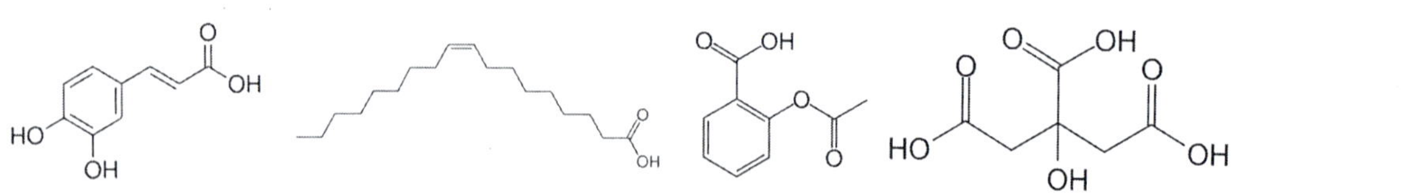

Caffeic acid ***Oleic acid*** ***Aspirin*** ***Citric acid***

2. Relating a general reaction mechanism to the specific example. Memorization of a reaction mechanism before its understanding is very tempting, but must be avoided at any costs. For instance, to apply the E2-mechanism shown below to a specific molecule, we need to decide which hydrogen can be abstracted.

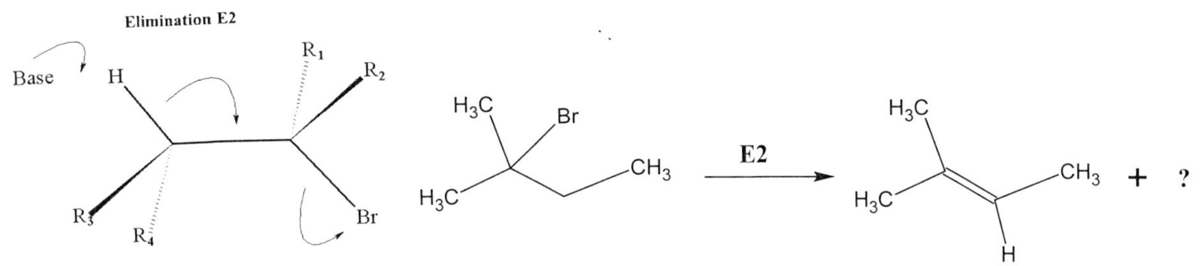

The reaction scheme shows that hydrogen must be attached to a carbon, which is next to the carbon losing the leaving group. The reaction scheme does not mean that the abstracted hydrogen must be at the left top part of the molecule. To practice, count how many hydrogens in the molecule of 2-bromo-2-methylbutaane are available for abstraction in an E2 reaction. If this number is not 8, make a molecular model and orient it in 8 possible ways to match the E2 reaction scheme. Now, you can draw the second product of dehydrobromination of 2-bromo-2-methylbutane. Make sure to not count same product twice.

3. Due to the electrolytic dissociation, solvation, and dimerization, reagents can be presented in a variety of ways. For instance, the catalyst in acid-catalyzed reactions can be presented as H_2SO_4, H^+, or H_3O^+, and for the reagent for hydroboration of alkenes, you can find formulas BH_3, B_2H_6, $BH_3 \cdot THF$ etc.

4. When we shift the acid-base equilibrium toward the conjugate base by increasing pH of solution above the acid's pK_a value, we need to realize, that we use an external acid or base to adjust the pH. Otherwise, it might be hard to understand, why a lower (more acidic) pH corresponds to a lower concentration of the conjugate base, rather than increases it due to dissociation that produces the conjugate base along with protons.

5. A set of curved arrows corresponds to only one step of the reaction mechanism. All arrows must be followed at the same time, and the next step requires a completely new set of curved arrows. Trying to write the structure of the step product by only one curved arrow from the set, or using one set of arrows to derive the product of a multistep reaction mechanism, are common errors.

6. Confusion between resonance structures and molecules at the equilibrium. Equilibrium happens between different molecules with different geometry and is usually shown by two oppositely pointed arrows. Resonance structures refer to the same molecule, so cannot be in an equilibrium. Relationship between resonance structures is shown by one double-speared arrow.

Equilibrium **Resonance**

Unfortunately the resonance arrow is used sometimes for chemical equilibria, which feeds the confusion. Oftentimes even instructors are using a term "stability of the resonance structure" actually thinking about its contribution to the compound's structure.

Pushing the limits

By drawing a structural formula, we assign a pair of electrons to each chemical bond and each non-bonding orbital, which is a necessary condition for building structural formulas. However, such distribution of electrons does not exactly correspond to the energy minimum, which makes most of organic chemistry an art of approximations. Making up for the deviations using resonance structures and other concepts of intramolecular interaction (inductive, resonance, and steric effects) can be taken to a quantitative level, using the computational method of natural orbitals, introduced by Foster and Weinhold in 1980. Lewis structures are not applicable to some particles, containing electrons on anti-bonding orbitals (He_2^+, O_2, excited and transition states of organic molecules). Fortunately, such problematic areas usually can be considered separately from the rest of the molecule with the additional information shown as dotted lines, multiplicity symbols, etc.). However, for the rest of the molecule, Lewis structures can still be used.

The next step of learning theory of acids and bases involves equilibria in non-aqueous solvents and the solvent's influence on acidity and basicity. The acid-base theory is resting on fast equilibria between particles with constant ability to donate or accept electron pairs. Naturally, when the proton exchange is slow, or particles drastically change their ability to donate or accept electrons in the process of the reaction, or depending on the reaction partner, the concept needs to be modified and expanded. Most common expansions are the concept of kinetic acidity, nucleophilicity, soft and hard acids and bases, and theories of ligand exchange around metals.

Practice problems

1. What is wrong about the following reaction scheme?

   $CH_3OH + HBr = CH_3Br + H_2O$

   A. The reaction scheme is correct

   B. It is not clear whether CH_3Br or H_2O is the final product

   C. It is not clear whether CH_3OH is consumed on the first step or produced on the second step.

   D. It is not clear whether CH_3Br is produced on the first step or consumed on the second step.

   E. It is not clear whether HBr is consumed on the first step or produced on the second step.

2. Which of the following equations is <u>not</u> balanced?

   A. $C_7H_{16} + 11O_2 = 7\ CO_2 + 8\ H_2O$

   B. $4\ Fe_3O_4 + O_2 = 6\ Fe_2O_3$

   C. $K_2CrO_4 + Pb(NO_3)_2 = 2\ KNO_3 + PbCrO_4$

   D. $2\ Al_2O_3 = 4\ Al + 3\ O_2$

   E. $3\ AgNO_3 + (NH_4)_3PO_4 = Ag_3PO_4 + NH_4NO_3$

3. In the reaction shown, how many moles of CH_2Cl_2 are produced from 10 moles of CH_4?

   $CH_4 + 2\ Cl_2 = CH_2Cl_2 + 2\ HCl$

   A. 10 **B.** 5 **C.** 20 **D.** 2 **E.** 2 HCl

4. What is the formal charge on carbon (marked with asterisk and has one unpaired electron) in the following particle?

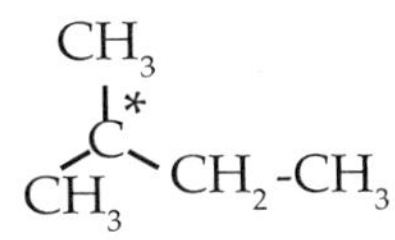

A. -2 **B.** -1 **C.** 0 **D.** 1 **E.** 2

5. What is the formal charge on carbon (marked with asterisk and has no unshared electrons) in the following particle?

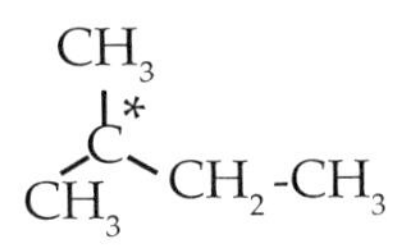

A. -2 **B.** -1 **C.** 0 **D.** 1 **E.** 2

6. What is the geometrical arrangement of substituents around the atom of carbon (marked with asterisk and has no unshared electrons) in the following particle?

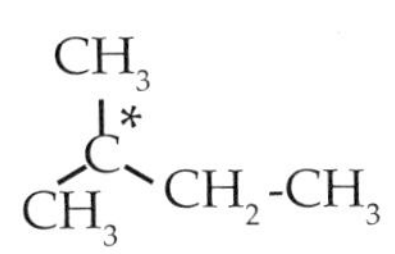

A. Trigonal pyramid **B.** Planar triangle **C.** Linear
D. Planar rectangle **E.** Tetrahedral

7. What is the geometrical arrangement of substituents around the atom of carbon of the CH_3-group in the following particle?

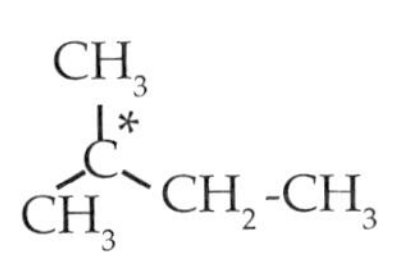

A. Trigonal pyramid **B.** Planar triangle **C.** Linear
D. Planar rectangle **E.** Tetrahedral

8. Which information about a reaction mechanism is given by the surface of potential energy?
A. Number of reaction steps **B.** The rate limiting step **C.** The heat effect of the reaction
D. The rate limiting step of the reverse reaction **E.** All of the above

9. What is true about the structures shown below?

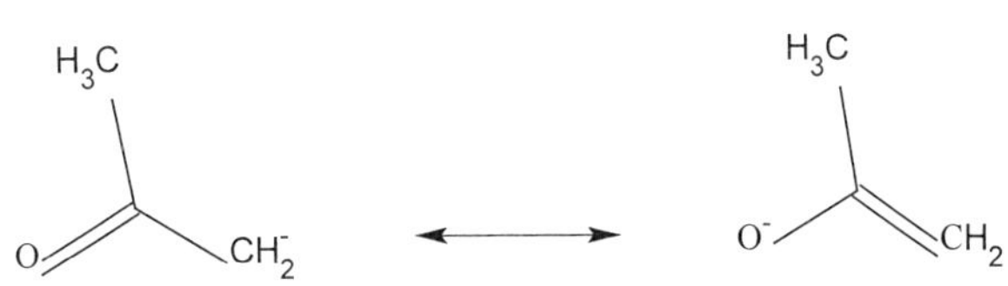

A. The two ions are in the equilibrium
B. The two ions have a different distribution of charges

C. The negative charge is distributed between two oxygen atoms of the same ion

D. The negative charge is distributed between one oxygen atom and one carbon atom of the same ion

E. There is not enough information to answer this question

10. Which of the following hybridization types is described in the following statement? "One p-orbital of oxygen is excluded from participation in the hybridization"

A. sp **B.** sp^2 **C.** sp^3 **D.** sp^5 **E.** no hybridization is considered

11. Which transformation includes the following reaction step?

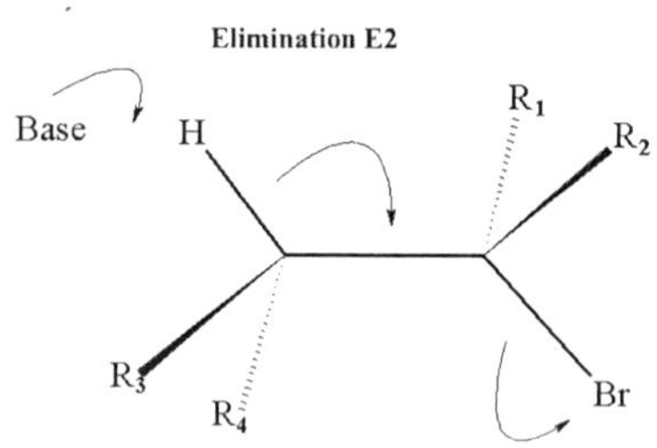

A. Second step of alkene hydrobromination

B. Second step of alkene hydration

C. Conversion of alkyl halides to alkenes

D. The shared step of alkene hydrobromination and alkene hydration

E. First step of alkene bromination

12. What information about a chemical reaction step is provided by a curved arrow notation mechanism?

A. Heat effect, number of steps, reaction rates for all steps

B. Shape of the intermediate on each step

C. Redistribution of electrons and charges between the nuclei

D. Molecular coefficients for all reactants and products

E. Interaction of orbitals on each step

13. What happens in the following reaction step?

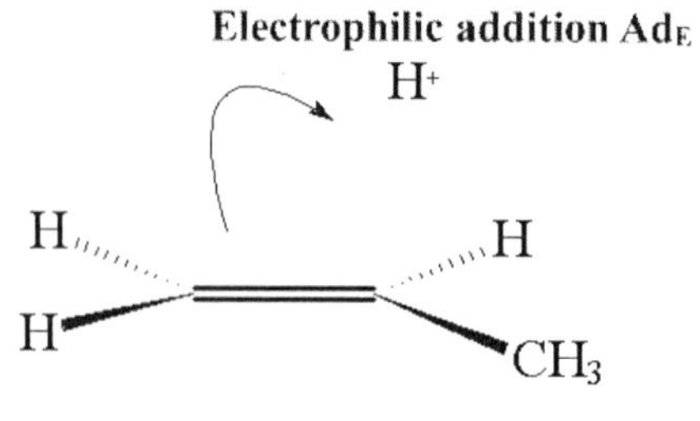

A. Second step of alkene hydrobromination

B. Second step of alkene hydration

C. Conversion of alkyl halides to alkenes

D. The shared step of alkene hydrobromination and alkene hydration

E. First step of alkene bromination

14. What happens in the following reaction step?

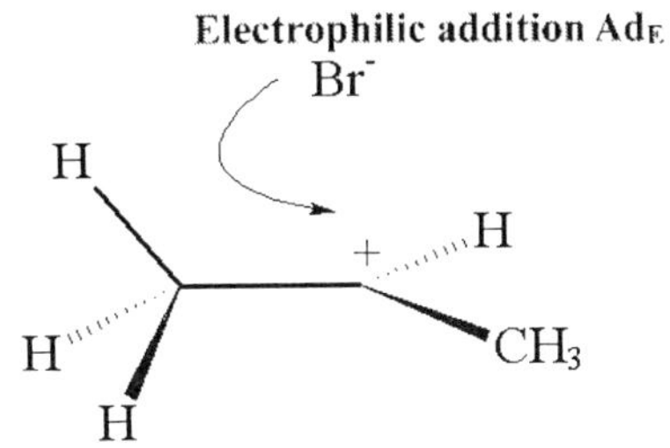

Second step starts

A. Second step of alkene hydrobromination

B. Second step of alkene hydration

C. Conversion of alkyl halides to alkenes

D. The shared step of alkene hydrobromination and alkene hydration

E. First step of alkene bromination

15. Use the following table to answer this question.

Acid	pK_a
RCOOH	5
CH_3OH (methanol)	15
NH_3	35
HF	3.2
HCN	9.4

What of the following reagents can convert methanol to its conjugate base in the highest yield?

A. $RCOO^-$ B. NH_2^- C. F^- D. CN^- E. CH_3O^-

16. Which of the following acids is the weakest listed?

A. RCOOH B. NH_3 C. HF D. HCN E. CH_3OH

17. At which pH value concentration of RCOOH exceeds concentration of its conjugate base in an aqueous solution?

A. 5 B. 15 C. 35 D. 3.2 E. 9.4

18. What of the following is true about the following reaction?

$RCOO^- + H_2O = RCOOH + OH^-$

A. The reaction does not take place

B. The equilibrium is shifted to the left

C. The equilibrium is shifted to the right

D. The reaction is fast

E. The reaction is slow

19. What of the following reagents can generate the highest concentration of OH^- from water?

A. $RCOO^-$ **B.** NH_2^- **C.** F^- **D.** CN^- **E.** HCN

20. Which of the following reaction steps leads to the formation of a new carbon-carbon double bond?

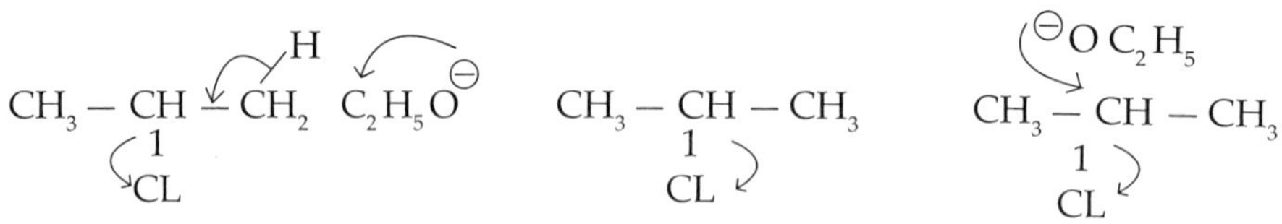

A **B** **C**

A. Structure A **B.** Structure B **C.** Structure C

D. Structures A and B **E.** Neither structure

21. Which of the following reaction steps changes the formal charge on carbon?

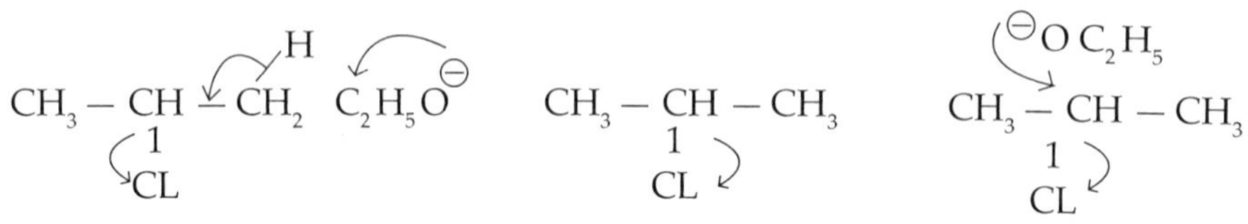

A **B** **C**

A. Reaction A **B.** Reaction B **C.** Reaction C

D. Reactions A and B **E.** Neither reaction

Making connections

The concept of Lewis structures justifies writing organic formulas and, therefore, is of paramount importance for the whole of organic chemistry. Knowledge of electronegativity and resonance structures allows to qualitatively assess energies of intermediates and predict the preferred reaction pathway. Acid-bases equilibria are must-have tools for understanding which particles are actually present in the reaction mixture and, therefore, may become reactants. The curved arrow notation based on the concept of Lewis structures is the only way to present a reaction mechanism and truly constitutes the centerpiece of the language of organic chemistry.

Chapter 2

Alkanes

Alkanes are made only of atoms of carbon and hydrogens bound by only single bonds. Therefore, each carbon is connected to four other atoms – hydrogens or carbons. The covalent bonds to those four neighbors point to the vertexes of a tetrahedron, which is best described in terms of sp^3 hybridization. The complete electron octet and lack of lone electron pairs on each carbon (as opposed to nitrogen, boron, oxygen, hydrogen, and most of other atoms) allows for the formation of stable carbon-carbon chains, which accounts for the unprecedented variety of carbon-containing compounds and uniqueness of organic chemistry. Depending on the information to be conveyed to the reader, there are several ways to draw organic structure with different levels of detalization. It is common to use different levels of detalization for different parts of molecule, for instance, showing more detail around a reaction center.

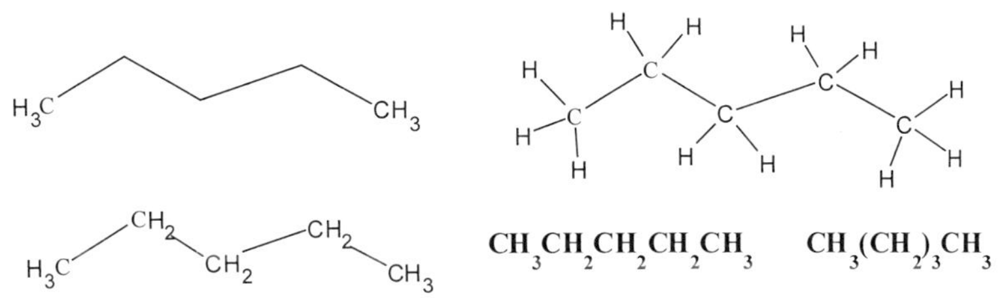

Pentane

The most detailed structure of pentane shows that each atom of carbon is bonded with two atoms of hydrogen. In addition, there are two more hydrogens (one on the left side, and one on the right side of the formula. In other words, number of hydrogens = (number of carbons)·2 +2, which means that all alkanes can be presented by one general molecular formula: C_nH_{2n+2}, which is termed *homologous formula* of alkanes. Now we can write a molecular formula for any alkane if we know the number of carbons in the molecule. It is also worth to note that the homologous formula holds true for any isomer, because all isomers have same number of atoms of each type.

Isomers of alkanes are characterized by different extents of branching of their carbon chain. In order to build an isomer of pentane, we need to remove a CH_2-group of the molecule and insert it between two neighboring atoms elsewhere in the molecule. This operation changes the order in which atoms are connected. Therefore, *isomers of alkanes are structural (constitutional) isomers.*

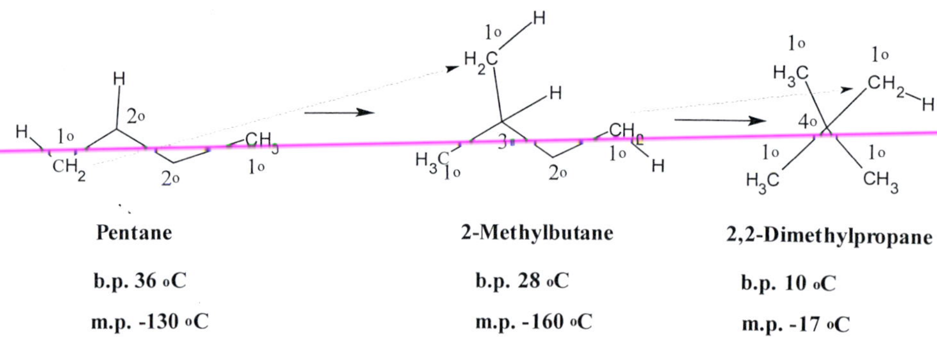

In compounds with more than one carbon, carbons are divided into four groups: *primary (1°, bonded to one carbon), secondary (2°, bonded to two carbons), tertiary (3°, bonded to three carbons), and quaternary (4°, bonded to four carbons).* Type of the carbon plays an important role in its chemistry.

For example, the structures of 2-methylbutane and 2,2-dimethylpropane are derived from the structure of pentane by removing one or two CH_2-groups from the carbon chain and inserting them in C-H bonds in the middle of the molecule.

In order to distinguish isomers and present their chemical identity in a non-graphical form, every organic compound is assigned with an individual name independent of the way its structure is drawn. Relatively simple compounds can be named by the free version of *ChemSketch* available from the internet. This software is an excellent tool for building 3D-organic structures, rapid optimization their geometry, and deriving chemical names. For the detailed set of rules for naming compound (also called *nomenclature*), we refer the reader to any organic chemistry textbook or online manuals. Students need to understand two basic rules of chemical nomenclature of any organic compound:

- The structure to be named is divided to two parts: 1. Principal carbon chain, which is usually the longest or cyclic one. Sometimes, the *principal chain* is required to contain certain functional groups such as multiple bonds. 2. The rest of the molecule is not included into the principal chain is called *substituents.*

- *Division of a molecule into the principal chain and substituents has nothing to do with chemical properties.*

In the example of 2-methylbutane, four carbons constitute the principal chain, and the CH_3 (methyl group) is considered as a substituent bonded to the carbon #2 of the principal chain. Please note, that the molecule of 2-methylbutane has *two chemically identical methyl groups* (bonded to the CH-group). One of those methyl groups is considered as a *part of the principal chain*, and another one – as a *substituent.*

Due to the little difference in the electronegativity of carbon and hydrogen (small dipole moment of C-H bonds), and the tetrahedral geometry around carbons (cancelling of C-H dipole vectors), alkanes are the least polar organic compounds. As a consequence, their intermolecular interaction completely relies on the weak Wan-der-Waals attraction, which increases as the molecular surface area increases and the molecules have a higher chance of bumping into each other. *The stronger the intermolecular interaction,* the less likely molecules jump from the liquid to the gas phase, which *increases the boiling point.* Larger molecules with more carbons usually have a higher surface area and, therefore, higher boiling points. Among the molecules of similar size, more branched molecules have lower surface areas than less branched ones, because their shape is closer to the sphere, which is known to have the least possible surface area. The tendency of more branched alkane to have lower boiling points is clearly seen on the figure above.

The trend of melting points is not that straightforward, because melting points depend not only on the intermolecular interaction, but also on how well molecules can be packed into the crystal lattice.

2,2-Dimethylpropane with compact sphere-like and easily packable molecules has a significantly higher melting point than its isomers.

Due to their low polarity, alkanes are poorly soluble in water and are considered *hydrophobic*. On the other hand, there are readily soluble in non-polar liquids (such as other alkanes) and are considered *lipophilic*. The term "hydrophobic" may mislead to a conclusion that there is repulsion between molecules of alkanes and molecules of water. As a matter of fact, molecules of alkanes and water attract each other. However, a molecule of alkane placed in an aqueous environment disrupts strong interactions (dipole-dipole and hydrogen bonds) between molecules of water. This energy loss is not compensated by the weak attraction between water and alkanes, which destabilizes the system and drives the alkane molecule back to the bulk of the non-polar alkane.

Currently, most of alkanes are produced by distillation of dehydrated and desalted petroleum followed by desulfurization. The alternative catalytic synthesis of alkanes from carbon monoxide and hydrogen (Fischer-Tropsch process) will be gradually replacing natural sources of alkanes as we are running out of petroleum. The use of alkanes in industry can be divided into three groups:

1. Use of alkanes as is (lubricating oils, solvents).
2. Use of alkanes as fuels (combustion to carbon dioxide and water).
3. Use as feedstock for organic synthesis (chemical modification).

Processing of petroleum always includes two major steps – cracking and reforming.

In the process of *cracking* performed at high temperatures with or without a catalyst, the *C-C bonds break* producing free radicals, which further exchange atoms of hydrogens giving rise to shorter alkanes and hydrocarbons with C=C (double) bond (alkenes).

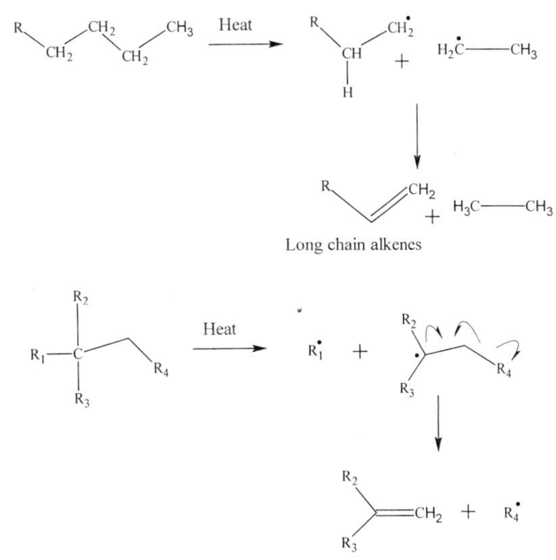

Long chain alkenes

Short chain alkenes

As opposed to cracking that "cracks" C-C bonds of the carbon skeleton, *reforming* targets C-H bonds. Reforming is mostly a sequence of catalytic eliminations and additions of hydrogen (dehydrogenation and hydrogenation), leading to the most stable mixture of alkanes and aromatic compounds, which does not contain alkenes. Depending on the catalyst, both cracking and reforming can be accompanied by isomerization of alkanes to their more stable branched isomers.

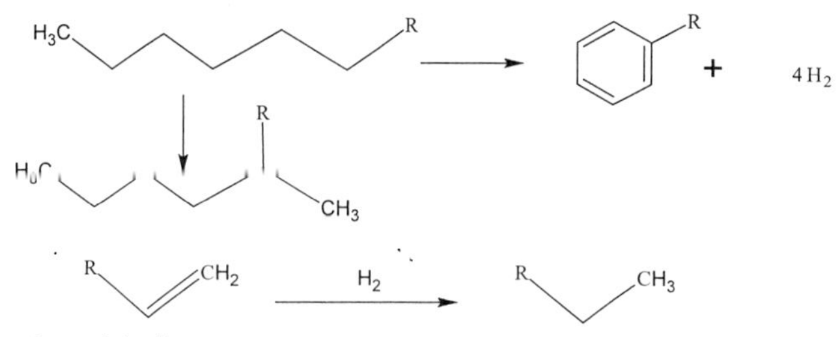

The following chapters will discuss why alkenes are less stable than alkanes especially on air, and why aromatic compounds are so stable that shift the equilibria of reforming toward their formation. For now, it is important to understand that *presence of alkenes in fuels is undesirable*, because it decreases stability and, therefore, quality of fuels.

Another important characteristic of fuels is octane rating, which shows the fuel's resistance to detonation, which is premature ignition of a fuel-air mixture before the spark plug fires. Combustion is a free-radical reaction initiated by a certain concentration of reactive free radicals. *Premature ignition is caused by excessive concentration of free radicals, which is linked to the amount of secondary carbons in the fuel*, able to generate secondary free radicals. *Secondary free radicals*, which are more stable than primary, and less stable than tertiary, are stable enough to be formed in the compressed fuel-air mixture, and unstable enough to react further and *continue the free-radical chain of combustion*. Therefore, more branched alkanes usually have higher octane rating than their less branched isomers. In the industry, heptane (octane rating is said to be 0) and 2,2,4- trimethylpentane (isooctane, octane rating is said to be 100) are used as references for measuring octane rating of fuels.

Heptane

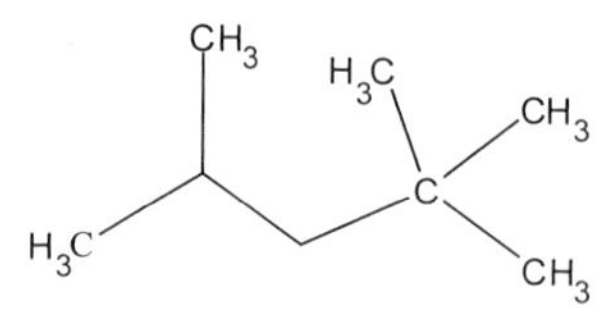

2,2,4-Trimethylpentane (isooctane)

A sample of fuel is assigned the octane rating of X, if it has same resistance to detonation, as a mixture of X% isooctane and (100-X)% of heptane.

Distillation of petroleum produces a mixture of alkanes (straight-run gasoline) with the octane rating below the industrial standards. Besides of fuels, vacuum distillation of petroleum produces lubricating oils and paraffin, which can be used as themselves or go to the step of cracking together with the undistilled residue. The additional mixture of hydrocarbons produced by cracking is even worse fuel that the straight-run gasoline because of the presence of alkenes reducing the fuel's stability. Therefore, reforming is an essential step in the production of alkene-free high octane fuel. Besides of providing fuels, providing cracking is the main source of short-chain alkenes (ethylene, propylene, and isobutylene), and reforming is the main industrial source of benzene and alkylbenzenes.

From now on, I highly recommend using some type of molecular models – physical (commercial sets or gum ball and toothpick models will work equally efficiently) or computer models (ChemSketch or any other molecular modeling software). At his point it is important to understand that carbon chains in alkanes are very flexible, so parts of the molecules can easily rotate around C-C single bonds making up various molecular shapes, which rapidly equilibrate at room temperature. *Rapidly interconverting molecular shapes are termed conformations* and as opposed to isomers cannot by isolated as individual compounds. Now, let's look at the simplest case – rotation around the C-C bond of ethane. There are two major ways to represent a molecular fragment in 3D: to look across three C-C bonds aligned in the same plane (wedge formula shown on the bottom of the figure below) or to look along a selected C-C bond (Newman projection shown on the top of the figure below).

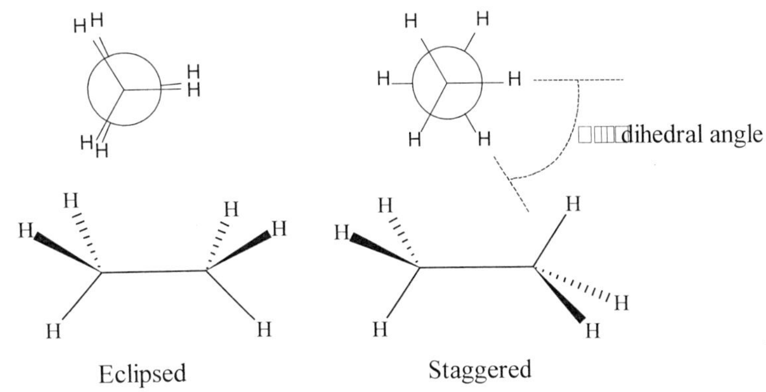

In a wedge formula, we align at least two bonds in the plane of the drawing and denote them by solid lines. Then, bonds that point toward us are denoted by solid wedges, and bond pointing away from us are denoted by dashed wedges. In a Newman projection, the closest to us carbon of the bond we are looking at is represented by a dot with three adjacent bonds, and the second, further carbon of the bond we are looking at is represented by a circle with three adjacent bonds.

Now, we need to define a parameter to describe conformational changes. We can do it by aligning three bonds in the plane of drawing and defining one plane by bond 1 and 2, and the second plane – by bonds 2 and 3. The angle between those two planes is termed *dihedral* or *torsional angle*. Changing of the dihedral angle corresponds to the internal rotation of the left part of the molecule with respect to the right part of the molecule around the C-C bond, which is example of a conformational change.

The Newman projection shows that in one conformation, the bonds at the farther carbon are eclipsed by the bonds at the closest carbon, which is why this shape with the dihedral angle = 0° is naturally termed "*eclipsed conformation*". In the conformation with the dihedral angle 60°, we see alternating bonds from the closest and the farthest carbon. This bond arrangement corresponds to the local minimum of potential energy, so at a sufficiently low temperature, the molecule gets stuck in this shape, which is termed "*staggered conformation*". In the eclipsed conformation of ethane, repulsion of the C-H bonds from different carbons (example of *steric repulsion*) contributes to increasing energy of the conformation, bringing it to an unstable point of local maximum. In the molecules more complex than ethane, presence of different groups attached to each carbons may lead to more than one staggered or eclipsed conformations.

Cycloalkanes. Removing one atom of hydrogen from each of non-adjacent carbons of an alkane and connecting those carbons together makes a cyclic compound – *cycloalkane*. Naturally, cycloalkanes contain two less hydrogens than alkane, which corresponds to the homologous formula C_nH_{2n}.

Cyclization imposes two major restrictions on the molecular geometry:

1. Bond angles may have to deviate from the value of about 109° characteristic for the symmetric tetrahedral geometry. Thus, in the molecule of cyclopropane, the C-C-C angle has to be just 60°. The difference of 49° is termed *angular strain*, which increases potential energy of the molecule, and usually makes it less stable. For larger cycles, the angular strain tends to relax, so potential energy increases from cyclopentane through methylcyclobutane to dimethylcyclopropanes.

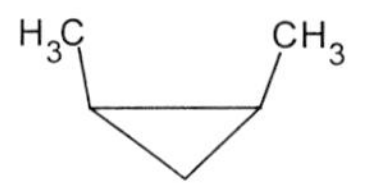

cis-1,2-Dimethylcyclopropane

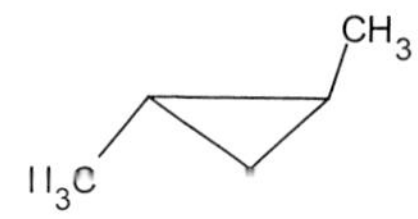

trans-1,2-Dimethylcyclopropane

2. Rotation around C-C bonds is restricted as a condition of keeping the cycle from opening. While by looking at each C-C bond we can still distinguish staggered and eclipsed conformations, the molecule as a whole has a unique set of conformations, depending on the cycle size. This restriction also brings about a very important phenomenon – stereoisomerism. *In stereoisomers, atoms are connected in the same order (same connectivity), but differently positioned in space.* If we draw a plane as close as possible to all carbons of a cycle, then due to the restricted rotation, atoms above and below that plane cannot switch around. This type of isomerism is termed *cis- trans-isomerism*. In the following example, *cis*-1,2-dimethylcyclopropane has a higher energy than its *trans*-counterpart due to the steric repulsion of more closely positioned methyl groups.

What can go wrong?

1. The major challenges are visualization of conformational changes and ability to navigate through a variety of structural formulas representing the same compound. At this point, students need to get a feel of the organic structure by practicing with physical or computer molecular models, and avoiding at all costs the temptation to use photographic memory for organic structures, or even worse – organic reactions.

2. Often students are making unjustified connections between properties of branched and unbranched hydrocarbons. Branching of the carbon chain leads to a slight decrease of the molecule's energy, boiling point, and increases octane rating. However, all those three characteristics are not directly linked, and should not be confused.

Pushing the limits

1. Expanding the scope of compounds and conditions under consideration reveals that there is no distinct borderline between conformations and stereoisomers. The rate of conformational changes around a C-C bond can be decreased by reducing temperature and placing bulky substituents at the bond to the extent that conformations can be isolated as individual compounds (isomers). Fine tuning the barriers of conformational changes has spawned a new direction in the materials science – molecular gears and molecular switches.

2. From general chemistry we have learnt the rule – "like dissolves like", which is based on dividing compounds on polar and nonpolar – hydrophilic and hydrophobic, which are also considered lipophilic. However, the world of molecules is not that "black and white". Fluorinated hydrocarbons with their unique bond polarization tend to be both hydrophobic *and* lipophobic, bringing about a group of very useful inert materials – Teflon and its analogues.

Practice problems

1. Which of the following conformations contains Br in the anti-staggered position to the CH_2OH group?

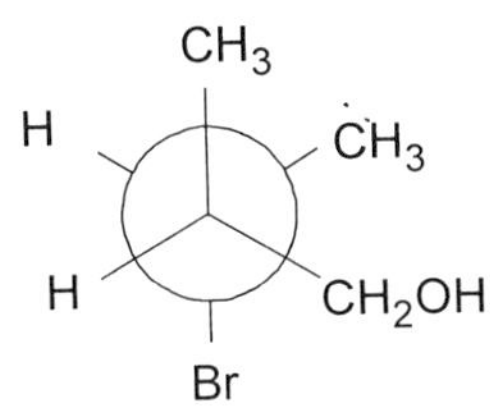

A

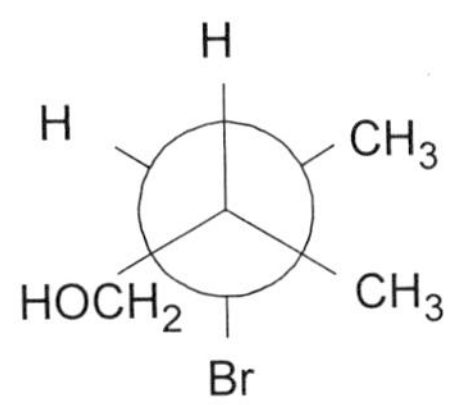

B

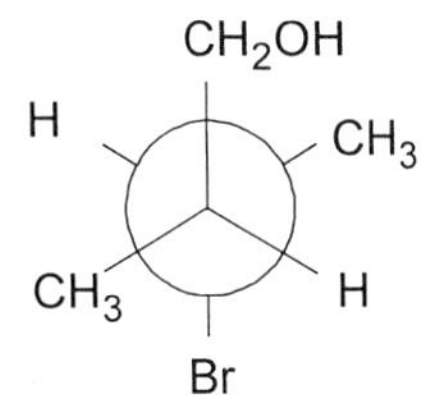

C

A. Conformation A **B.** Conformation B **C.** Conformation C **D.** Neither **E.** No preference

2. Which compound, I or II, contains a principal carbon chain with 8 carbons?

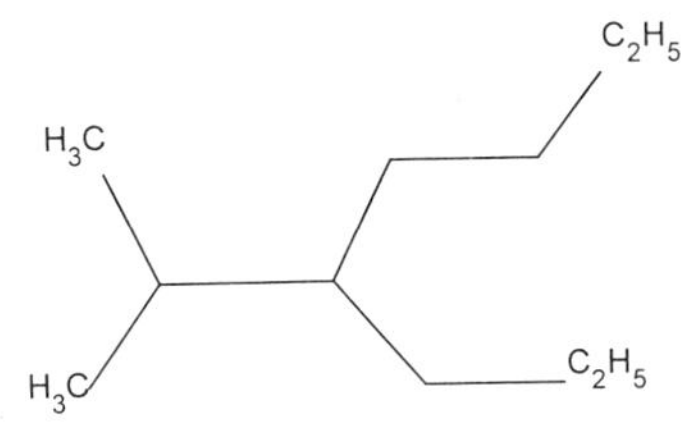

I

II

A. I **B.** II **C.** Both I and II **D.** It depends on the boiling point

E. It depends on which of two alternative approaches is used to select the principal chain

3. Which of the following compounds belongs to the class of alkanes?

A. $C_{10}H_{22}$ **B.** C_6H_{12} **C.** C_6H_6 **D.** $C_5H_{12}O$ **E.** $C_5H_{10}O$

4. Which of the following compounds is a secondary alcohol with a tertiary carbon?

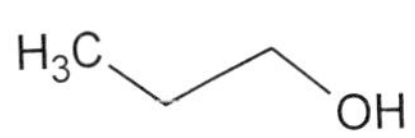

A

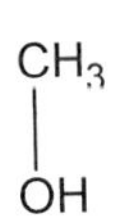

B

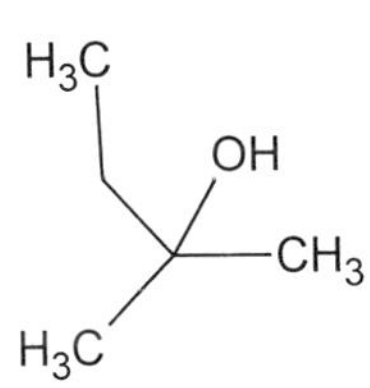

C

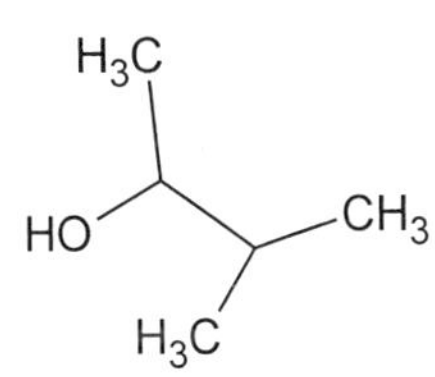

D

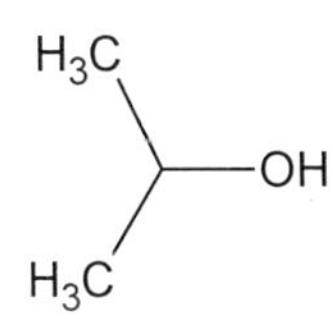

E

5. Which of the following chemical names is wrong?

A. 2-Methylhexane **B.** 3-Methylhexane **C.** 1-Methylhexane **D.** 2,3-Dimethylhexane

E. 2,2-Dimethylhexane

6. Which of the following is true?

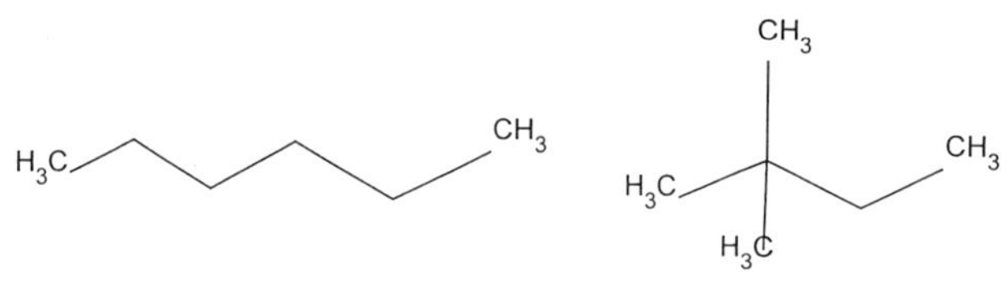

I II

A. **I** has a higher boiling point, than **II** because of larger surface area of **I**

B. **I** has a higher boiling point, than **II** because of lower surface area of **I**

C. **I** has a lower boiling point, than **II** because of larger surface area of **I**

D. **I** has a lower boiling point, than **II** because of lower surface area of **I**

E. There is not enough information to answer this question

7. Which does the following formula show about the position of the OH- group?

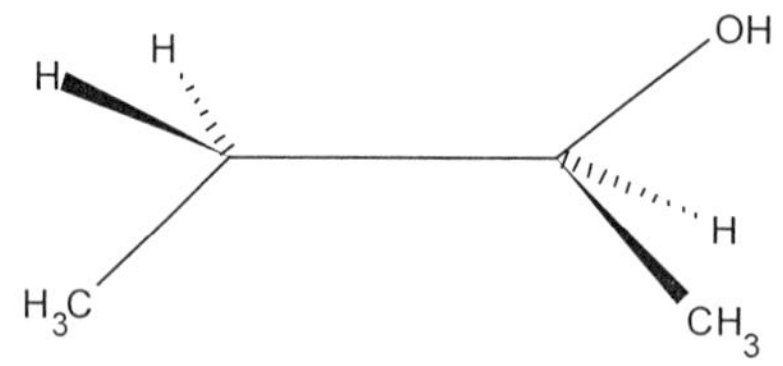

A. It is located in the plane of the drawing

B. It points behind the plane of drawing?

C. It points above the plane of the drawing

D. It is not connected to the carbon principal chain

E. It has priority in numbering

8. What is true about the following two conformations?

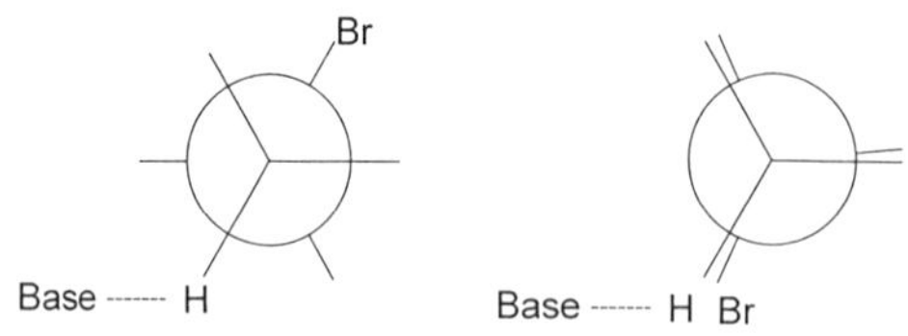

I II

A. **I** is more stable, than **II**, because **I** is an eclipsed conformation

B. **I** is less stable, than **II**, because **I** is a staggered conformation

C. **I** is more stable, than **II**, because **I** is a staggered conformation

D. **I** is less stable, than **II**, because **I** is an eclipsed conformation

E. There is not enough information to answer this question

9. Which of the following molecules has the overall dipole moment equal zero?

A. HBr **B.** Dichloromethane **C.** Chloromethane **D.** Tetrachloromethane

E. Chloroethane

10. Which parameter changes most with conformational changes?

A. Bond angle **B.** Torsion (dihedral) angle **C.** Bond length **D.** Hybridization of carbon

E. Bond polarity

Making connections

- While the chapter of alkanes is placed after acids and bases, it makes sense to discuss them with students right after introduction of Lewis structures – as their first examples.

- Conformational changes play the key role in flexibility and physical properties of polymers. When a piece of stretched rubber bounces back, the polymeric molecules contract due to conversion of more stable anti-staggered conformations to less stable gauche-staggered conformation, spontaneously going uphill of the potential energy driven by enthropy. Holding a rubber band in a stretched position, we can tangibly feel the force of enthropy. A significant portion of biochemistry is driven by conformational changes (enzymatic catalysis, bioaffinity). Formation of wrong conformations is the molecular basis of many diseases such as Alzheimer's and prion infections.

Chapter 3

Alkenes and Alkynes

Alkenes are hydrocarbons with one carbon-carbon double bond (C=C). Both carbons connected with the only C=C bond are bonded with three atoms. Therefore, those carbons have the planar geometry, they are sp^2-hybridized, and have a p-orbital.

So far, we know how to determine geometry around atoms. Now, we need to look at the geometry of bonds.

- (Sigma)-bonds are symmetrical with respect to rotation around the bond.

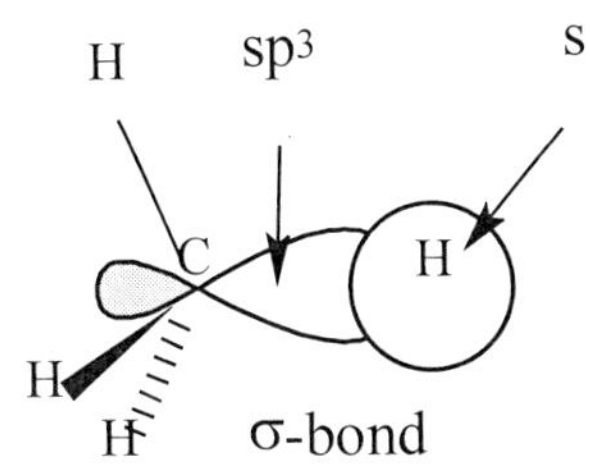

Consequence: Rotation of fragments around the bond does not break the bond.

- **(Pi)-bonds** are non-symmetrical with respect to rotation around the bond.

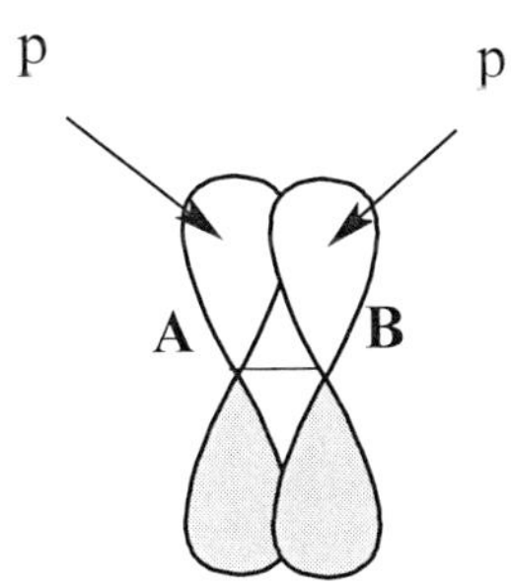

Consequences: 1. Rotation of fragments around the bond breaks the bond. 2. The π-bonds are weaker, than the σ-bonds, due to a weaker orbital overlap.

Together, **σ- and π- bonds form a double bond,** so two components of the C=C bond are not equal.

Summary of the structure of alkenes:

1. The valence shell of the atom of carbon has one s-orbital and three p-orbitals.

2. When the carbon is not bonded by any π-bonds, all s- and p-orbitals are involved in the formation of σ-bonds through the sp3-hybrid orbitals.

3. When the carbon is bonded by one p-bond, one p-orbital participates in the formation of this π-bond and the remaining one s-orbital and two p-orbitals are involved in the formation of σ-bonds through the sp2-hybrid orbitals.

4. The presence of sp^2-hybridized carbons (hosts of sp^2-orbitals) is characteristic for alkenes.

Due to the planar geometry of the double bond, two substituents can be located at either the same side of the double bond, or at the opposite sides of the double bond.

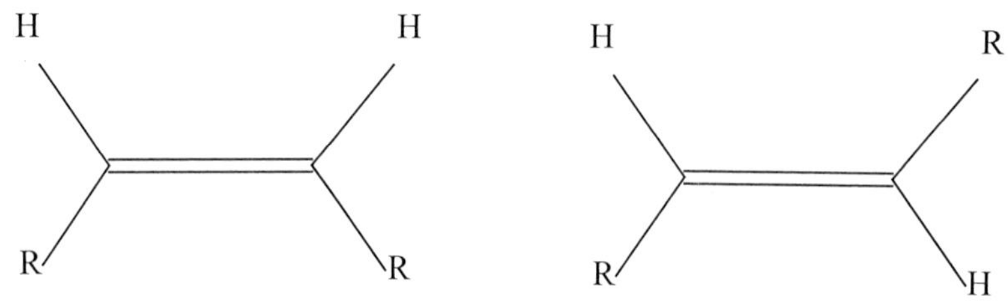

cis-isomer **trans**-isomer

Note: cis-trans-isomerism is impossible if at least one carbon at the double bond has two identical substituents.

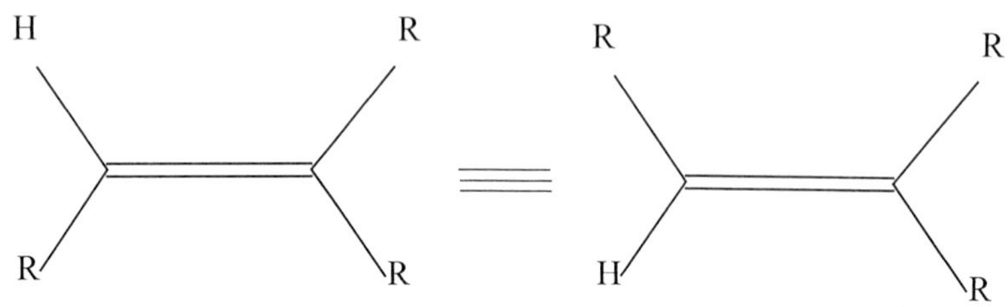

identical

Besides of the cis-trans-isomerism, there are four more differences between the nomenclature of alkenes and nomenclature of alkanes:

1. Replace "-ane" with "-ene"

2. The principal carbon chain must contain the double bond

3. While numbering the principal chain, the double bond must have the lowest number

Note: Number of the double bond is the number of the first carbon encountered

4. In the chemical name, indicate position of the double bond

Example

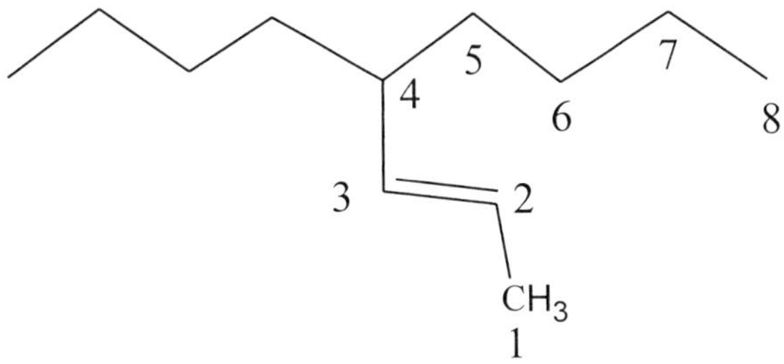

trans-4-Butyl-2-octene

The second consequence of the π-bond structure "*The π-bonds are weaker, than the π-bonds, due to a weaker orbital overlap*" leads to a much higher reactivity of C=C bonds versus carbon-carbon single bonds.

The elevated reactivity of C=C bonds best manifests itself in the reactions of *addition,* during which two things happen:

1. The C=C bond breaks.
2. The freed valences of both carbons from the former C=C bond connect to other atoms.

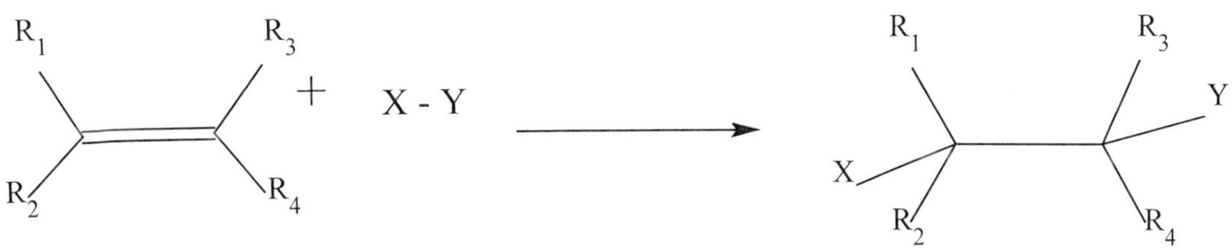

The simplest examples of addition is hydrogenation – addition of hydrogen, which requires a catalyst.

$$H_2C{=}CH_2 + H_2 \rightleftharpoons H_3C{-}CH_3$$

Catalyst = Pt or Ni or other catalysts

A very common mechanism of addition to C=C bonds is electrophilic addition that starts from withdrawing an electron pair from the π-component of the C=C bond, and generating a cation.

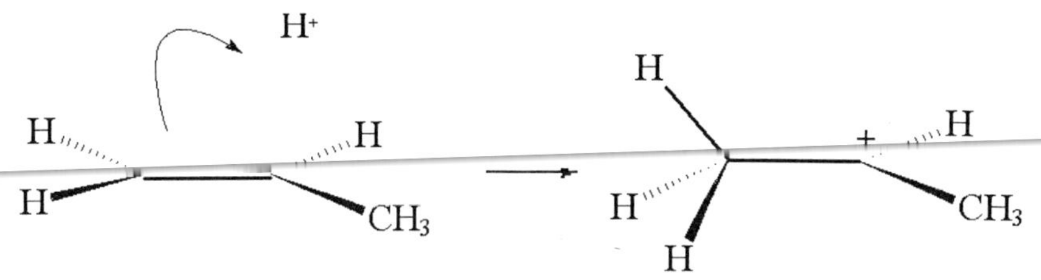

The particle withdrawing an electron pair from the C=C is a Lewis acid by definition. However, the acid-base theory is based on the equilibrium, but we need to focus on the reaction rate of electrophilic addition. That is why we need another concept to refer to that particle. That concept is "*Electrophilicity*".

Electrophiles accept an electron pair just like Lewis acids. Electrophiles are considered with a focus on the reaction rate, while Lewis acids are considered with a focus on the equilibrium. *Stronger Lewis acids shift the equilibrium more. Stronger electrophiles react faster.*

On the second step, the carbocation rapidly reacts with a particle, donating a pair of electrons. In the reaction of hydrobromination, it is Br^-.

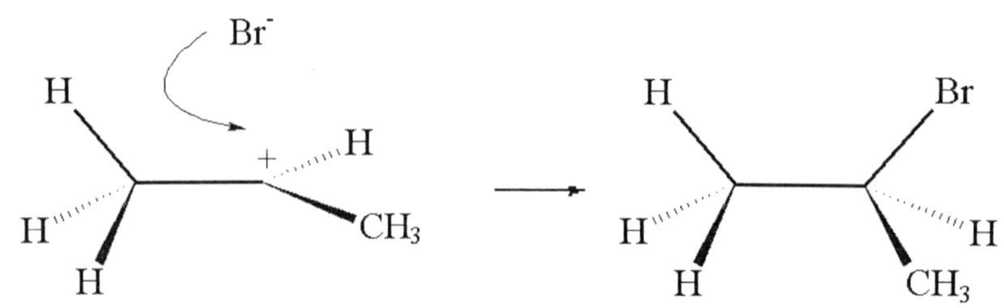

The particle donating an electron pair to the carbocation is a Lewis base by definition. However, the acid-base theory is based on the equilibrium, but sometimes we need to focus on the reaction rate. That is why we need another concept to refer to that particle. That concept is "*Nucleophilicity*".

Nucleophiles donate an electron pair just like Lewis bases. Nucleophiles are considered with a focus on the reaction rate, while Lewis bases are considered with a focus on the equilibrium. The exceptions are *Lewis bases,* donating a pair of electrons to a hydrogen atom, or, in other words, acting as *Bronsted bases.* Such bases are not considered as nucleophiles regardless of the reaction rate, because the proton transfers are described very well in terms of the acid-base theory. *Stronger Lewis bases shift the equilibrium more. Stronger nucleophiles react faster.*

We will discuss the concept of nucleophilicity in detail in the chapter "Halides, Alcohols, and Ethers".

Here is the overall mechanism of hydrohalogenation: http://ochem.orgfree.com/AdE.html

The Markovnikov rule

It is very common, that addition to alkenes can lead to at least two products. For instance, for hydrohalogenation, to which of two carbons will the hydrogen preferably add? Obviously, the halogen will add to the remaining carbon. The choice of the major product of addition should be made with the consideration of the ***Markovnikov rule: hydrogen adds to the carbon with more attached hydrogens, and the other group adds to the carbon with fewer attached hydrogens.***

The Markovnikov rule works for alkenes, if the reaction proceeds by the two-step electrophilic mechanism exemplified above. This happens, because the first step produces a more substituted carbocation, which is more stable due to its positive charge partially spread to the substituent.

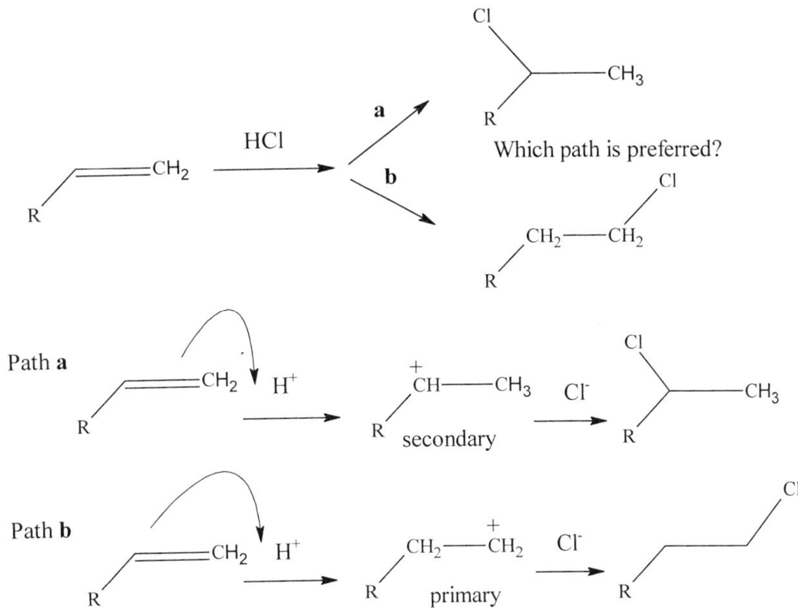

The path a is preferred according to Markovnikov rule.

Another example of addition by Markovnikov rule is acid-catalyzed hydration of alkenes.

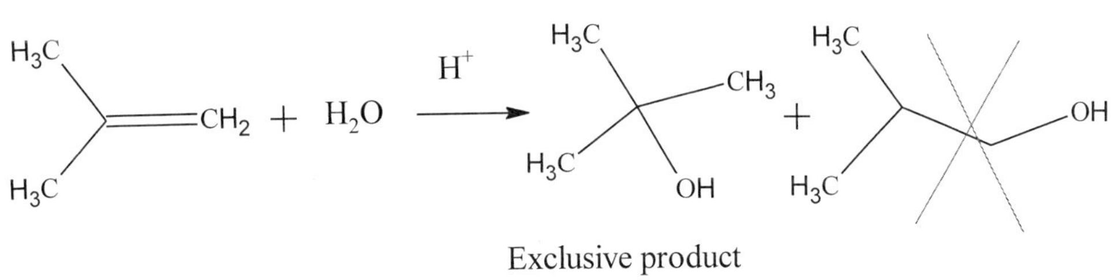

The Markovnikov rule may or may not work, if the reaction proceeds by a different mechanism or if the compound with the double bond contains additional structural moieties, such as double bonds or halogenated groups.

Examples of addition against Markovnikov rule:

1. *Hydrobromination in the presence of peroxides,* because the reaction switches to the free radical mechanism.

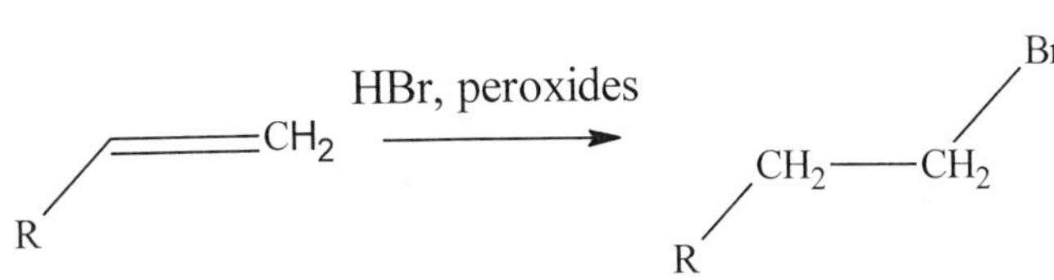

2. *Hydroboration, followed by oxidation (overall – anti-Markovnikov hydration of C=C bonds)*, because the B-H bonds adds to a C=C bond synchronously without the involvement of a carbocation.

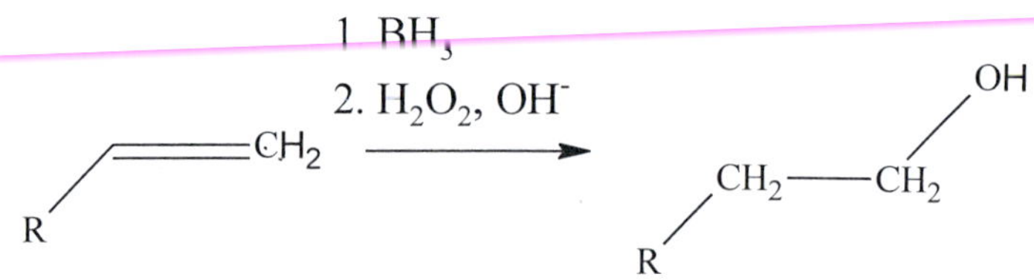

3. Hydrohalogenation of an unsaturated ketone, because protonation proceeds to the atom of oxygen, which is more basic than the C=C bond.

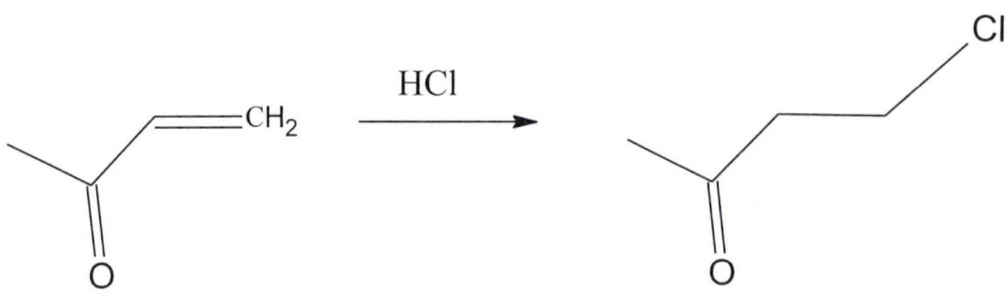

When the C=C bond is attacked by an electrophile larger than hydrogen, the positive charge in the intermediate spreads mostly over three atoms, which shifts the reaction path from a carbocation to a three-membered cyclic cation, as it takes place in the reaction of halogenation.

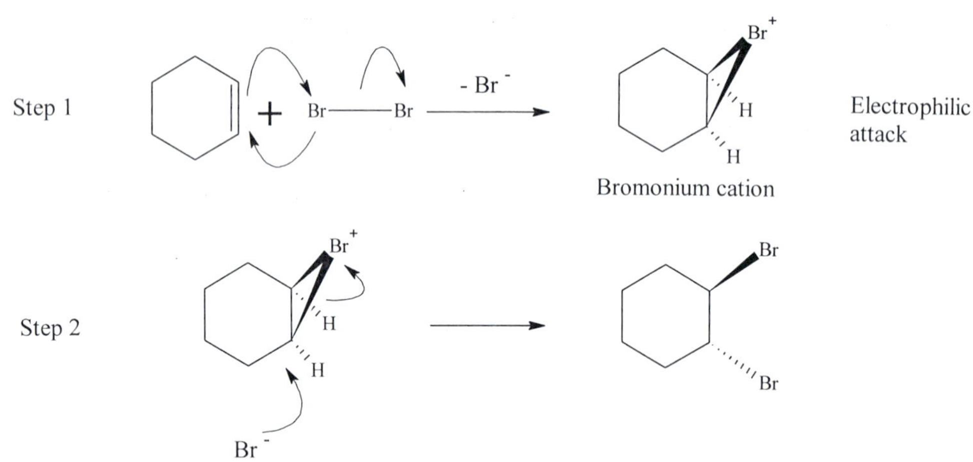

Polymerization of alkenes

Under certain conditions, the π-electrons of the C=C bonds can bind those carbons with their counterparts from different molecules. This is the process of polymerization, which results in the formation of large molecules with a repeating unit. Those molecules are termed *polymers*, making up many synthetic materials such as plastics.

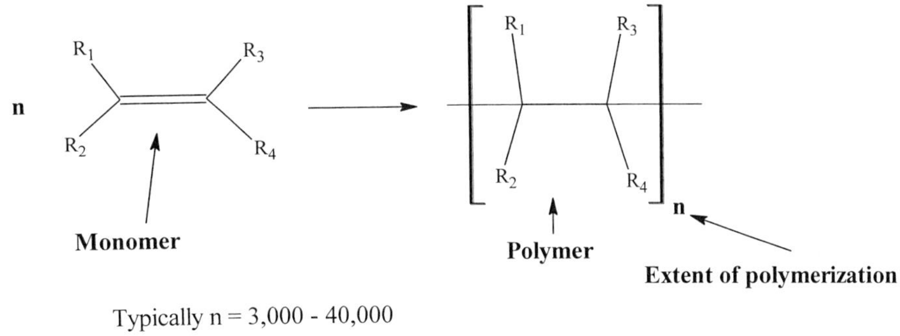

To correctly draw the structure of a polymer, based on the structure of the monomer:

1. Identify two carbons, connected by the C=C bond. Only those carbons will become a part of the polymer chain.

2. If necessary, re-draw the structure of the monomer to remove all substituents from the horizontal line and avoid the common mistake of including those substituents to the polymer chain.

3. Break the π-component of the C=C bond.

4. The freed valences of each carbon will hold together the monomer units in the polymer chain.

5. Show the monomer units by the brackets, and extend the freed valances over the brackets.

To correctly draw the structure of the monomer, based on the structure of a polymer:

1. Remove the brackets.

2. Connect the free carbon valences to a C=C bond.

Alkynes

Alkynes are hydrocarbons with one triple carbon-carbon bond, formed from one σ- and two π-components.

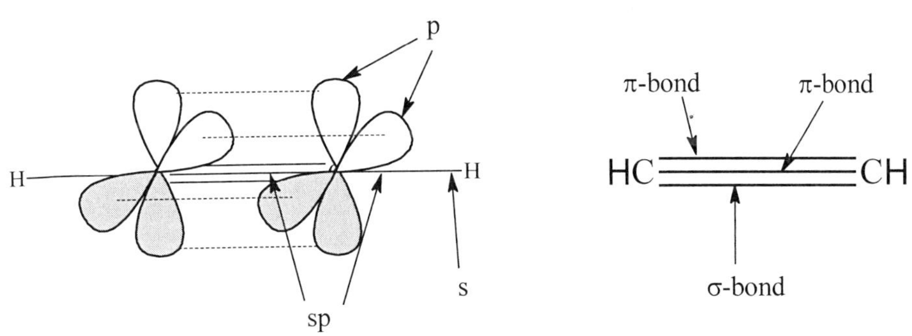

Now we need to choose a hybridization to describe systems, containing a triple bond. The valence shell of the atom of carbon has one s-orbital and three p-orbitals. When the carbon is bonded by two π-bonds, each of two p-orbitals participates in the formation of this π- bond, and the remaining one s-orbital and one p-orbital are equally involved in the formation of σ-bonds, giving rise to the sp-hybridization state.

The presence of sp-hybridized carbons is characteristic for alkynes.

The increased contribution of the lower in energy s-orbital of carbon to the σ-bonds increases electronegativity of the sp-carbon so much, that it exceeds electronegativity of nitrogen. Therefore, the terminal hydrogen (attached to the carbon-carbon triple bond) is more acidic than ammonia. Therefore, terminal alkynes can be significantly deprotonated with $NaNH_2$, which generates an acetylide – a strong nucleophile, whose application in organic synthesis will be discussed later with the topics of nucleophilic substitution and carbonyl compounds.

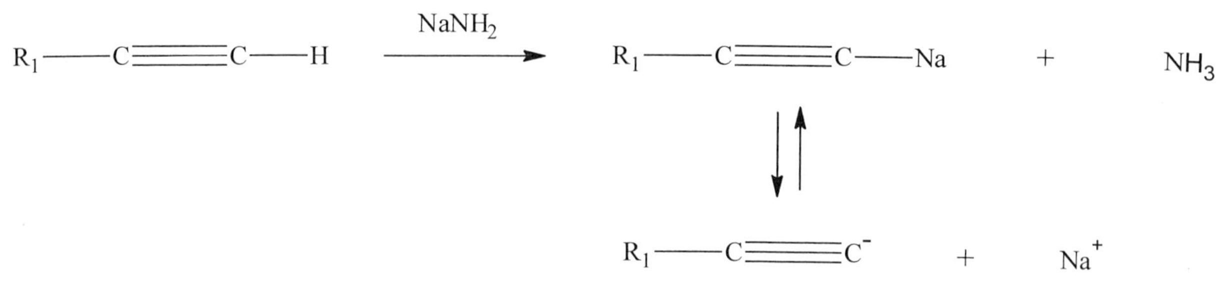

An acetylide (a strong nucleophile)

The nomenclature of alkynes is very similar to those of alkenes. However, due to the linear geometry at the sp-hybridized carbon, there can be no cis- or trans-configurations around triple carbon-carbon bonds.

Just like alkenes, alkynes undergo electrophilic addition. However, there are significant differences:

1. Alkynes are less reactive with electrophiles, because the reaction has to proceed through either unstable vinyl cation, or unstable three-membered ring with a double bond. Sometimes, the reaction requires a catalysts, which shifts the reaction to a more favorable pathway. For instance, hydrohalogenation is usually performed in the presence of quaternanry ammonium salts.

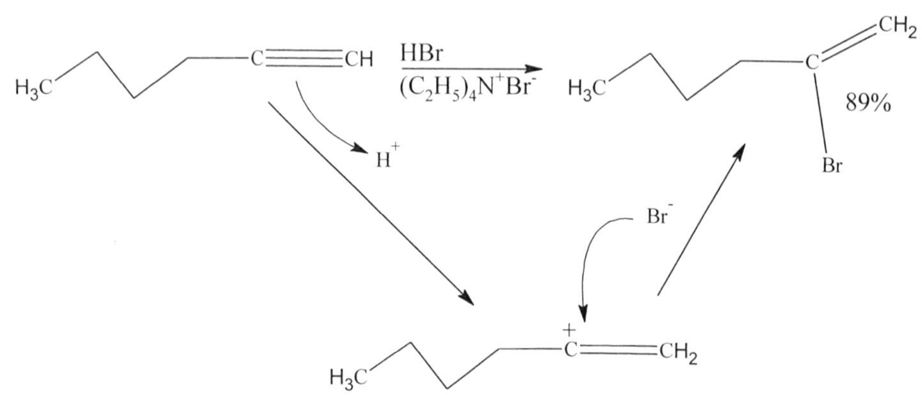

The intermediate vinyl carbocation is much less stable, than alkyl carbocations, which slows down the reaction

2. The product of electrophilic addition to alkynes has a C=C double bond. Therefore, the product may undergo addition with another equivalent of the electrophile, or rearrange. For instance, the product of hydrohalogenation reacts with excess of hydrogen halide.

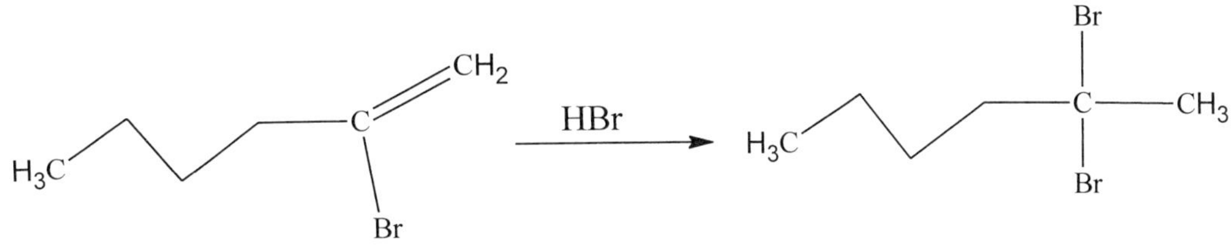

A unique reaction is hydration of alkynes that needs to be catalyzed by Hg^{2+}/H^+

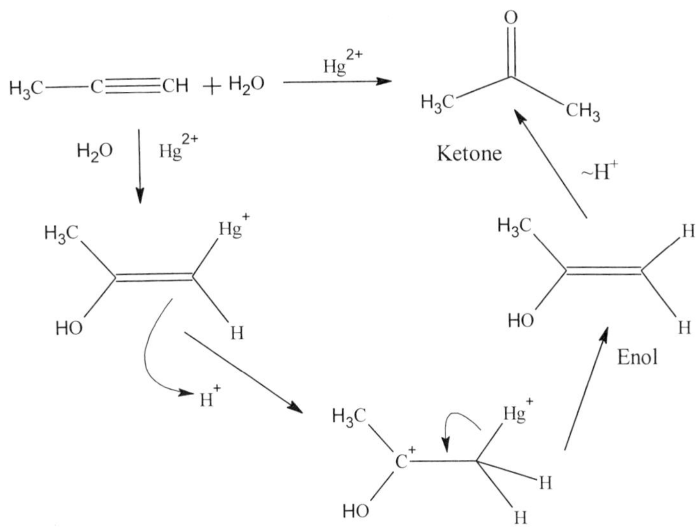

The product of water addition contains the -OH (hydroxy) - group next to a C=C double bond. This type of compounds is termed "Enolates", which are usually unstable, and rearrange to an aldehyde or a ketone.

The Markovnikov rule works for alkynes in a same way it works for alkenes, including the conditions when the addition proceeds against the rule. When we use the rule for hydration of alkynes, it should be applied to the intermediate enol (the product of addition before it rearranges).

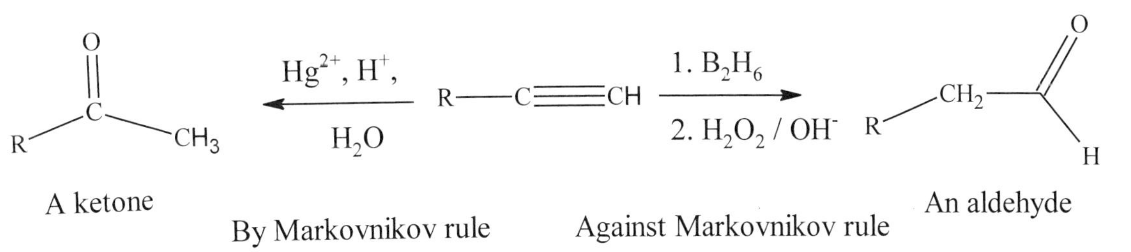

Hydrogenation of alkynes

The triple carbon-carbon bond is shorter than the C=C double bond, and better matches the short H-H bond in the molecule of H_2.

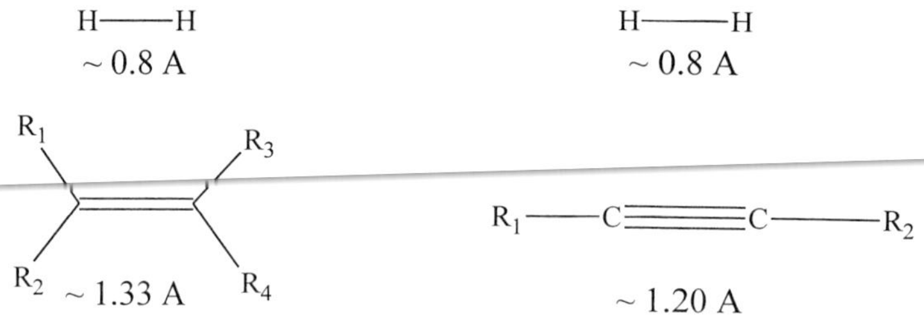

As a consequence, contrary to electrophilic addition, alkynes are more reactive towards catalytic hydrogenation than alkenes. Therefore, it is possible to design a catalyst that is active enough for hydrogenation of alkynes, but not active enough for hydrogenation of alkenes. This is the *Lindlar catalyst* (Pd partially poisoned with Pb or pyridine (Py)), which converts *alkynes to cis-alkenes.*

Trans-alkenes can be produced from alkynes under different conditions: reduction with alkali metals in liquid ammonia.

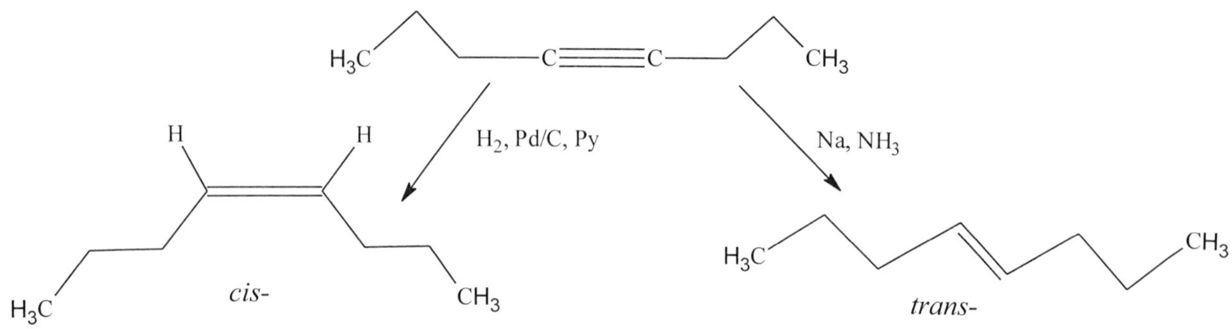

What can go wrong?

Here is the topic where we lose students, who did not practice enough with the concepts of chemical structure, hybridization, acids, and bases. Another advice would be to not start studying the section on alkynes before mastering their alkene counterparts.

Pushing the limits

Understanding the mechanism of electrophilic addition will allow you to predict its stereochemical outcome. The stereo-and regioselectivity of electrophilic addition may significantly differ, if intermediates are altered by reaction conditions. For instance, addition of HBr to alkenes proceeds against the Markovnikov rule, because the reaction shifts toward the free-radical path. Breaking the C=C bond in the excited state opens up a new realm of organic photochemistry. Plugging in the Hammond's postulate and shifting the focus from intermediate to the corresponding transition states will provide a much more accurate description of the reactions. The next step in studying the chemistry of alkynes is employing them as building blocks in multistep organic syntheses. Another recommended application of this material would be understanding of the partial hydrogenation of fats and looking into the different possible mechanisms of polymerization.

Practice problems

1. The following reaction step represents conversion of alkyl halides to alkenes. **Circle both carbons that will be connected with a C=C bond in the reaction product.**

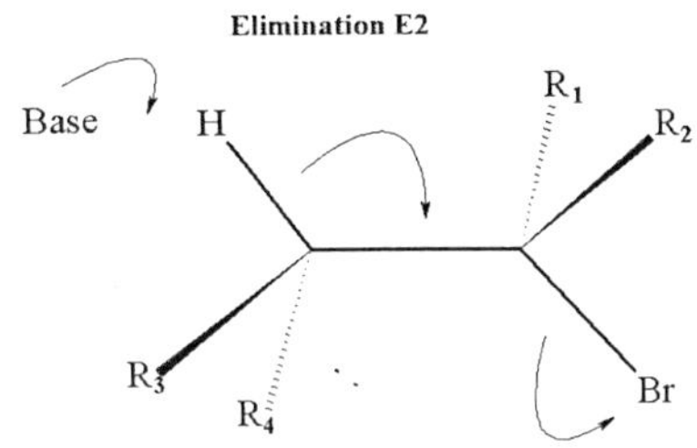

2. The following reaction step represents protonation of propene, leading to a secondary carbocation. **Circle the carbon that will have the formal positive charge in the resulting carbocation.**

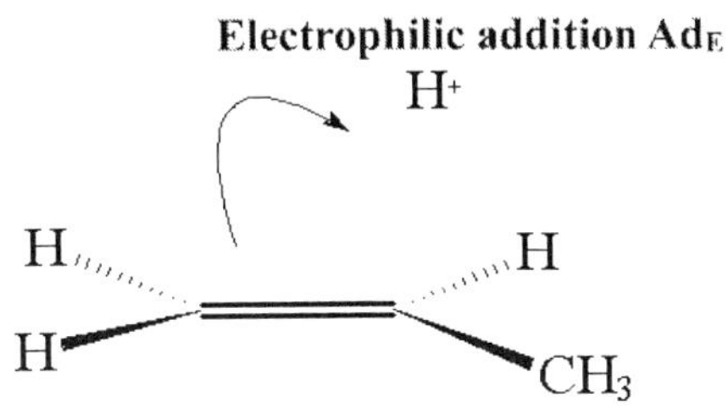

3. The following reaction step represents the second step of alkene hydrobromination. **Circle the carbon that forms a bond with Br.**

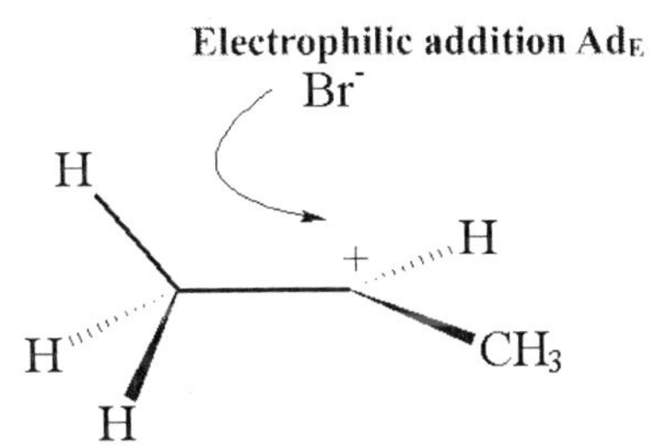

Second step starts

4. Polymerization of which compound will produce a polymer with the following structure?

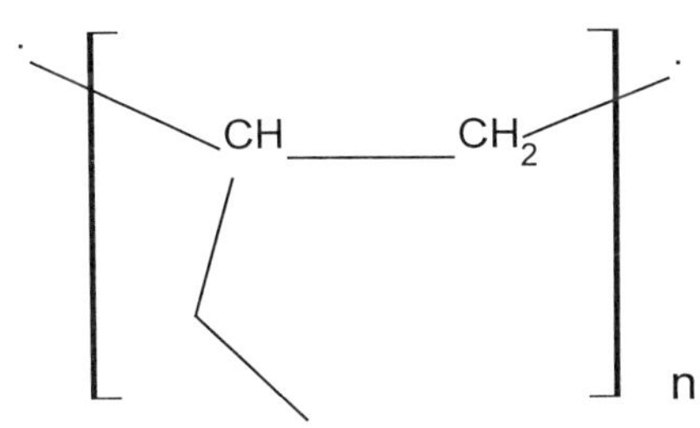

A. Ethene **B.** Propene **C.** Butane **D.** 1-Butene **E.** Propane

5. While cis-2-butene is a valid chemical name, there is no such compound as *cis*-2-butyne. **Why?**

A. Higher acidity of hydrogen at an sp-carbon than at an sp^2-carbon

B. Higher acidity of hydrogen at an sp-carbon than at an sp^3-carbon

C. Different geometry at the carbon with two electron domains and the carbon with three electron domains

D. Lower reactivity of alkynes than alkenes with electrophilic reagents

E. Higher electronegativity of sp-carbon, than electronegativity of nitrogen

6. In the structure of (1E)-1-bromo-1-chloro-2-methyl-1-butene (figure below), circle the highest in priority substituents, whose positions with respect to the C=C bond you use to assign a Z- or E-configuration to the compound.

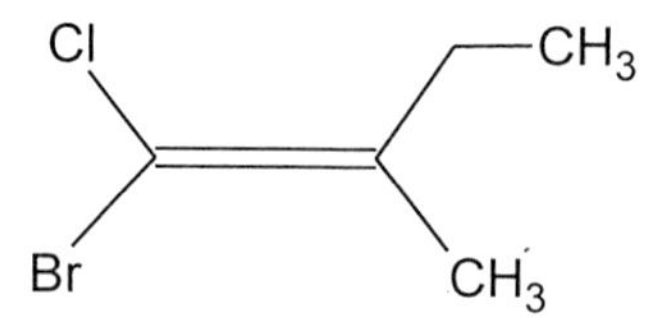

7. Circle the most acidic hydrogen in the following structure.

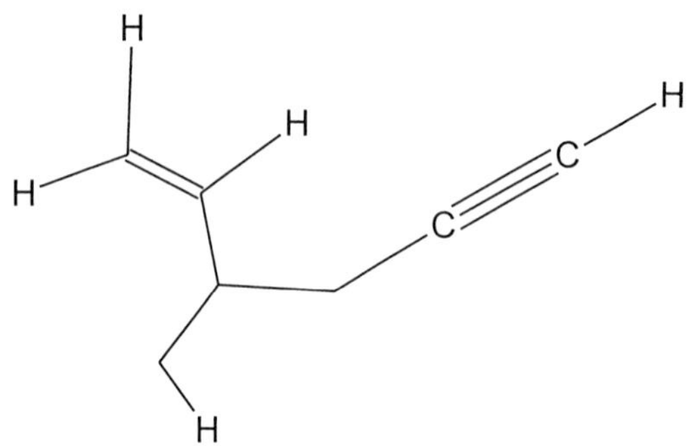

8. Rotation around a C=C bond in a molecule of butene is restricted. **What of the following is the consequence?**

A. Possibility of isomeric 1-butene and 2-methylpropene

B. Higher acidity of 1-butene, than butane

C. Possibility of isomeric *cis*-2-butene and *trans*-2-butene

D. Reactivity of butenes with bromine

E. Reactivity of butenes with water in the presence of acid

9. Interaction of which orbitals describes the component of the C=C bond, responsible for the existence of cis-trans-isomerism?

A. sp^3, sp^3. **B.** sp^2, sp^2. **C.** sp^2, p. **D.** sp, sp **E.** p, p

10. Arrange the following compound in the order of increasing acidity

A. NH_3 **B.** CH_4 **C.** C_2H_2 **D.** C_2H_4

Making connections

It has been already noted, that chemistry of alkenes and alkynes is so tightly connected with the chemical concepts discussed before, that only mastery of these concepts justifies the time invested to the studying this chapter. With the knowledge gained from the chapter, you will be ready to understand what happens when we place the C=C double bonds in chains and in cycles, master the concept of aromaticity, and appreciate how chemistry around the C=C bond enables our vision.

Chapter 4

Stereochemistry

Chemical reactions are often significantly impacted by the shapes of reactants in 3D. This aspect of chemical properties is the focus of stereochemistry – stereochemistry of compounds and stereochemistry of reactions.

For each atom in the molecule, we can list all atoms it is connected to, and by which type of bonds. For example:

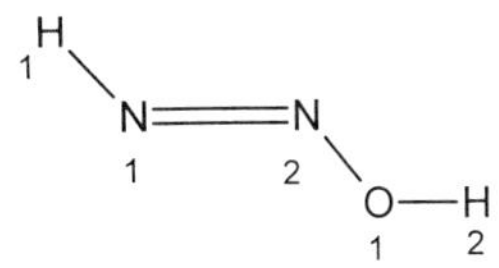

Hydrogen 1 is connected to nitrogen 1 with a single bond

Hydrogen 2 is connected to oxygen 1 with a single bond

Nitrogen 1 is connected to hydrogen 1 with a single bond

Nitrogen 1 is connected to nitrogen 2 with a double bond

Nitrogen 2 is connected to oxygen 1 with a single bond

Nitrogen 2 is connected to nitrogen 1 with a double bond

Oxygen 1 is connected to hydrogen 2 with a single bond

Oxygen 1 is connected to nitrogen 2 with a single bond

The list we just built, reflects *connectivity* of atoms.

If the connectivity is different for two isomers, the isomers are referred to as *structural (constitutional) isomers*. For example:

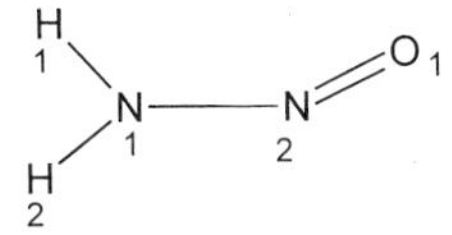

Hydrogen 1 is connected to nitrogen 1 with a single bond

Hydrogen 2 is connected to nitrogen 1 with a single bond

Nitrogen 1 is connected to hydrogen 1 with a single bond

Nitrogen 1 is connected to hydrogen 2 with a single bond

Nitrogen 1 is connected to nitrogen 2 with a single bond

Nitrogen 2 is connected to oxygen 1 with a double bond

Nitrogen 2 is connected to nitrogen 1 with a single bond

Oxygen 1 is connected to nitrogen 2 with a double bond

In fact, if even a single connectivity description is absent in the second isomer, those isomers are structural isomers. For instance, if there is a tertiary carbon in one compound, and there are no tertiary carbons in its isomer, those isomers are for sure structural isomers.

Sometimes, however, connectivity of all atoms in two isomers are same, but positions of some atoms in 3D are different. Such isomers are referred to as stereoisomers. For instance, the following compound is a stereoisomer to the original one. Please, compare their lists of connectivity.

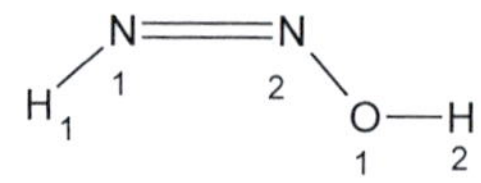

Hydrogen 1 is connected to nitrogen 1 with a single bond

Hydrogen 2 is connected to oxygen 1 with a single bond

Nitrogen 1 is connected to hydrogen 1 with a single bond

Nitrogen 1 is connected to nitrogen 2 with a double bond

Nitrogen 2 is connected to oxygen 1 with a single bond

Nitrogen 2 is connected to nitrogen 1 with a double bond

Oxygen 1 is connected to hydrogen 2 with a single bond

Oxygen 1 is connected to nitrogen 2 with a single bond

Cis- and *trans*-isomers that we learned in the chapter "Alkenes and Alkynes" is an example of stereoisomers.

Stereoisomers are divided into two types, according to the relationship with their mirror image.

Stereoisomers different from their mirror image are termed *enantiomers*.

Stereoisomers same as (also "congruent with", or "superposable with", or "superimposable with") their mirror image are termed *diastereomers*.

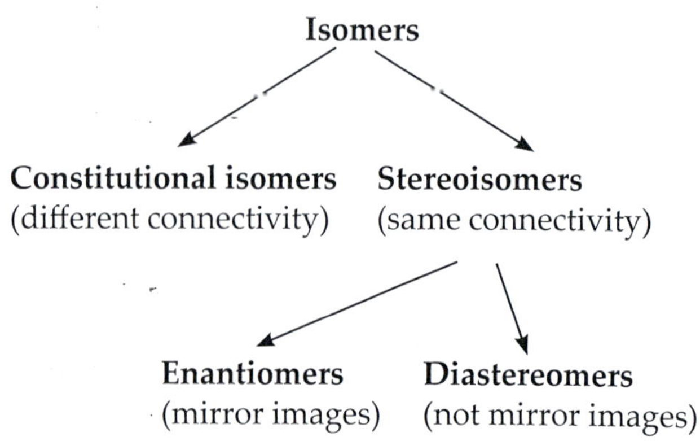

Cis- and *trans*-isomers is an example of diastereomers.

The nomenclature of stereoisomers utilizes the Cahn-Ingold-Prelog system, usually well explained in any organic chemistry textbook, where the reader is referred to. Here we will learn additional connections, which will facilitate mastery of the material.

The *E,Z-system* is used to assign a configuration to a C=C double bond.

1. Using the Cahn-Ingold-Prelog system, rank two groups at each carbon of the C=C bond.
2. From each pair, pick the highest-ranking group.
3. If the highest-ranking groups are on the same side of the C=C bond, the C=C bond configuration is "Z". Otherwise, it is "E".

However, not any C=C bond can be assigned a configuration. If at least one carbon at the C=C bond has two identical groups, we would not be able to rank them and, therefore, assign a configuration to the C=C bond. This gives you a simple guideline:

If you are not able to assign a configuration to a C=C bond, it does not have a configuration.

The ability of an object to be different from its mirror image (= to be *chiral*) is enabled by the presence of a chiral element. The most common chiral element for organic molecules is the chiral center, which is usually a chiral atom of carbon.

The *R,S-system* is used to assign a configuration to a chiral center.

1. Using the Cahn-Ingold-Prelog system, rank all for groups at the chiral center.
2. Rotate the molecule in 3D to point the lowest ranking group away from you.
3. Look if the priorities of the remaining three groups are decreasing clockwise or counterclockwise.
4. If the priorities are decreasing clockwise, the configuration is "R". Otherwise it is "S".

Compounds with one chiral center of the opposite configurations are always enantiomers, as shows the example below.

(R)- (S)-

The following rule will help you remember the step 4 for assigning a configuration to a chiral center.

When you draw a curve clockwise, it becomes a part of the letter "R", which corresponds to the "R"-configuration.

When you draw a curve counterclockwise, it becomes a part of the letter "S", which corresponds to the "S"-configuration as shown in the drawing below.

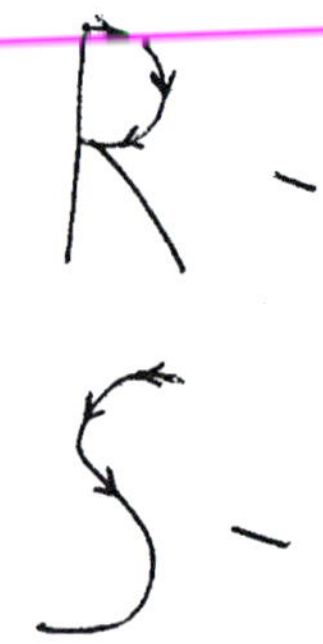

To assign a configuration, you must rank all four groups at the potential chiral center, which is impossible if any two groups are identical.

Therefore, you can use the same guideline that you applied to the E,Z-configurations:

If you are not able to assign a configuration to a potential chiral center, it is not a chiral center, and it does not have a configuration.

To understand stereochemistry of compounds with multiple chiral centers, it is essential to picture them in 3D. This is achieved by three types of stereochemical formulas:

1. Wedge-and-dash formulas
2. Newman projections
3. Fischer projections

For the detailed description of these formulas, the reader is referred to any organic chemistry textbook. Here, we will look at what exactly is described by each type of the stereochemical formulas, and will learn a few rules and shortcuts for handling them.

Both wedge-and-dash formulas and Newman projections show a *specific conformation* around a specific bond. While the Newman projection gives a view along the bond, the wedge-and-dash projection gives a different perspective. Both formulas also carry the information about the configurations of chiral centers. Since there is more than one way to look at an object, many wedge-and-dash formulas or Newman projections can correspond to the same conformation.

Fischer projections are specifically developed to represent compounds with multiple chiral centers. The Fischer projection is actually a "collage" of wedge-and-dash views of each carbon of interest, where the *wedges are always replaced with a solid horizontal line, and the dashes are always replaced with a solid vertical line.* Due to this convention, there is no need to use wedges and dashes in the Fischer projections. They are implied. Here is an example of D-erythrose:

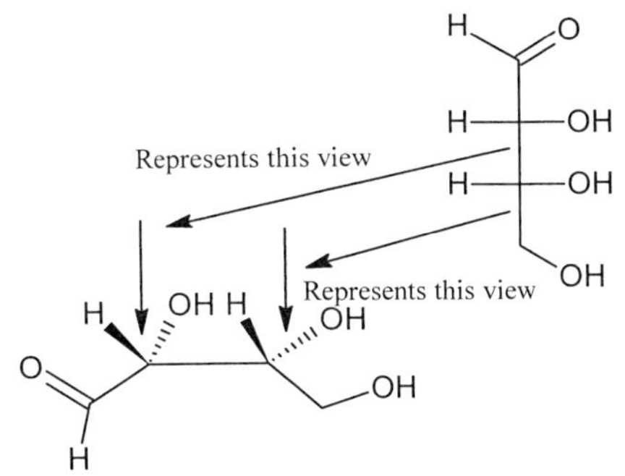

Please, note that to build a Fischer projection, we must use an arbitrary conformation of the molecule. Therefore, the Fischer projections do not represent any specific conformation, but they do represent configurations of all chiral centers. Sometimes, we can use both wedge-and-dash formulas, Newman projections, and Fischer projections to represent different parts of a complex molecule, depending on what we want to show.

Since we can look at a chiral center from different directions, one molecule can be represented by many Fischer projections. All those projections can be related to one another and interconverted by four simple rules:

1. *Rotation by 180º in the plane of the drawing gives the identical compound.*

(R)- (R)-

This rule works, because the vertical line remains vertical, and the horizontal line remains horizontal, so the implied wedges and dashes are not switched around.

2. *Rotation by 90º in the plane of drawing gives the opposite configuration.*

(R)- (S)-

This rule works, because the vertical line become horizontal, and the horizontal line becomes vertical, so the implied wedges and dashes are switched around. The Fischer projection becomes incorrect for the original configuration. However, there are only two options, so the resulting formula corresponds to the opposite configuration.

3. *Switching any two substituents gives the opposite enantiomer.*

(R)- (S)-

This rule works, because switching any two substituents produces a mirror image of the chiral center, which corresponds to the opposite configuration.

4. *Rotation of any three substituents gives the identical compound.*

(R)- (R)-

This rule works, because rotation of any three substituents represents rotation of the whole molecule around the bond connecting the chiral center with the forth substituent. This operation models the actual physical phenomenon.

For the formulas of molecules different only by the configurations of multiple chiral centers, here is how to determine their relationship:

1. *If configurations of <u>all</u> chiral centers are opposite, the formulas correspond to enantiomers.*

2. *If configurations of <u>all</u> chiral centers are same, both formulas correspond to the same compound.*

3. *If configurations of <u>some</u> chiral centers are opposite, and configurations of <u>some</u> chiral centers are same, the compounds are diastereomers.*

While the rule 3 is unambiguous, applying the rules 1 and 2 can be a bit tricky.

For instance, do the following formulas correspond to enantiomers or to the same compound?

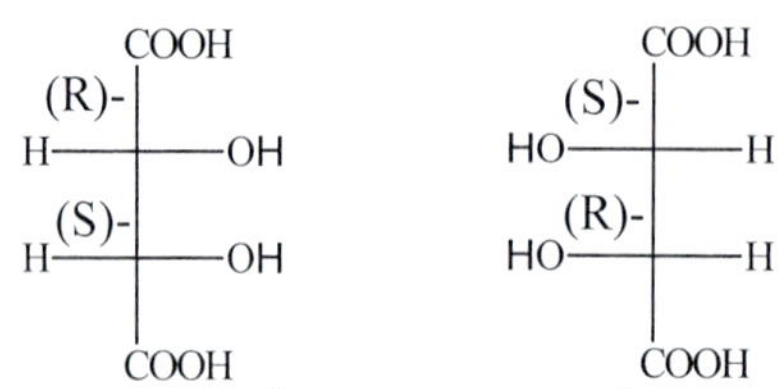

A meso-compound

If you compare the configurations of two centers at the top, and two centers at the bottom, all configurations are opposite. If you compare each center on the top with its counterpart on the bottom, all configurations are opposite. So, do the formulas correspond to an identical compound or to enantiomers? To answer this question, check for the presence of a plane of symmetry that may run between the chiral centers. For the above example, such plane does exist, which makes the compound achiral. Enantiomers cannot be achiral. Therefore, both formulas correspond to the same achiral compound, a *meso-compound*.

Generally speaking, *meso-compounds are achiral compounds with chiral centers.*

Now, do the following formulas correspond to enantiomers or to the same compound?

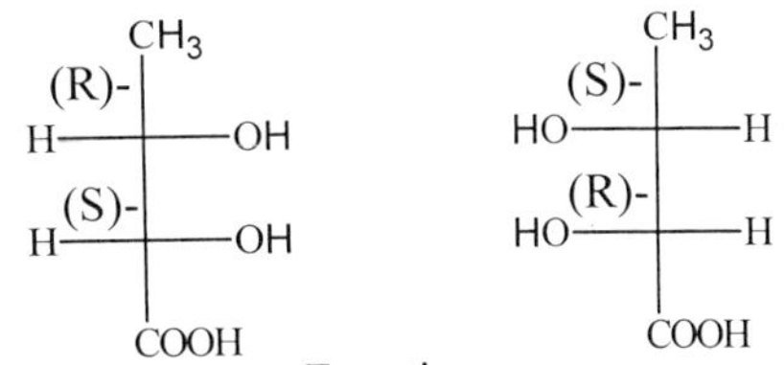

Enantiomers

To answer this question, check again for the presence of a plane of symmetry that may run between the chiral centers. For the above example, such plane does not exist, which makes the compounds chiral. Since all the chiral centers have opposite configurations, the formulas correspond to enantiomers.

The fact that there only are two possible configurations for a chiral center, gives you a couple more shortcuts.

1. In a Fischer projection, the lowest in priority group (hydrogen) most often points at you, so to assign the configuration, you would need to look at the chiral center from the opposite side, which takes some 3D-imagination. Instead, you can simply assign the configuration with the hydrogen pointing at you, at arrive to the wrong answer. Now, flip the answer, and you have the correct one.

2. If you need to draw a structure with a specific configuration, like (2S)-bromobutane, first draw a random configuration, and assign it. If it is the configuration you want, which is a 50/50 chance, you are done. If not – flip any two substituents, and the configuration flips with it.

Now, let us go over the main principle of chiral processes, such as synthesis, chromatography, extraction, catalysis, etc. All those processes must treat opposite enantiomers differently, which is not easy, because enantiomers have exactly same physical and chemical properties in an achiral environment. The overarching strategy is to reversibly convert enantiomers to diastereomers, which have different physical and chemical properties, and, therefore, they are much easier to separate. Here are some examples, whose specific applications can be found in any stereochemistry-related textbook:

1. *Separation of enantiomers.* Enantiomers are treated with a chiral reagent (for instance, a racemic mixture of acids forms a mixture of salts with a chiral amine). The products are diastereomeric, because the enantiomers introduce there opposite chiral centers, and the chiral reagent introduces same chiral centers.

   The diastereomeric products do not have to be isolated individual compounds. They can even be a weak associate, whose difference in energy may help discriminate the enantiomers.

2. *Enantioselective synthesis.* An achiral compound reacts with a chiral reagent, producing a new chiral center of either configuration (for instance, addition to a double bond). It results in diastereomeric transition states, which react further with a different rates, preferentially producing one diastereomer over its counterpart. At this point, it is *diastereoselective synthesis.* Removal of the chiral reagent from the preferred diastereomer leads to the preferred enantiomer.

3. *Chiral catalysis.* An achiral compound or intermediate reacts with a chiral catalyst, producing a new

chiral center of either configuration (for instance, enzymatic hydration of fumaric acid). It results in diastereomeric transition states, which react further with different rates, preferentially producing one enantiomer over its counterpart.

4. *Chiral chromatography.* Enantiomers are passed through a chiral sorbent. The complexes between the enantiomers and the chiral surface are diastereomeric, so the enantiomers are held by the sorbent with a different strength. As a result, one enantiomer moves through the chromatography column faster than its counterpart.

5. *Chiral extraction.* A solution of enantiomers is treated with a chiral extraction solvent. The solvates between the enantiomers and the chiral solvent are diastereomeric, so may have different energy contribution to the solubility. As a result, the enantiomers have different solubility in the extraction solvent, so one of them is extracted preferentially.

6. *Chiral crystallization.* A mixture of enantiomers is crystallized from a chiral solvent. The solvates between the enantiomers and the chiral solvent are diastereomeric, so may have different energy contribution to the solubility. As a result, the enantiomers have different solubility in the crystallization solvent, so one of them crystallizes preferentially.

7. *Seeded chiral crystallization.* A mixture of enantiomers in an achiral solvent is crystallized in the presence of a chiral crystal (seed), usually of one of the enantiomers to be separated. The complexes between the enantiomers and the surface of the chiral crystal are diastereomeric, so may have different energy contributions to the rate of the crystal growth. As a result, one of the enantiomers crystallizes preferentially.

8. *Chiral anything (generally speaking).* A mixture of enantiomers interacts with a chiral anything to produce two types of a diastereomeric something, which may have different energies, different properties, and, therefore, preferential behavior toward different enantiomers.

What can go wrong?

1. Identical groups that are presented on a structural formula differently, which can mask the presence of a plane of symmetry in the following compound:

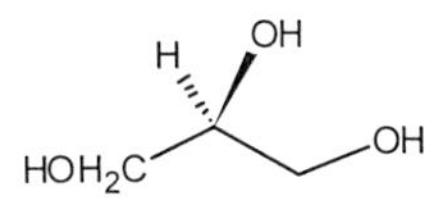

2. The same example also demonstrates that stereochemical formulas can be applied to non-chiral centers.

3. A drawing of a symmetric molecule does not have to be symmetric. It can make less obvious that the following structures are mirror images:

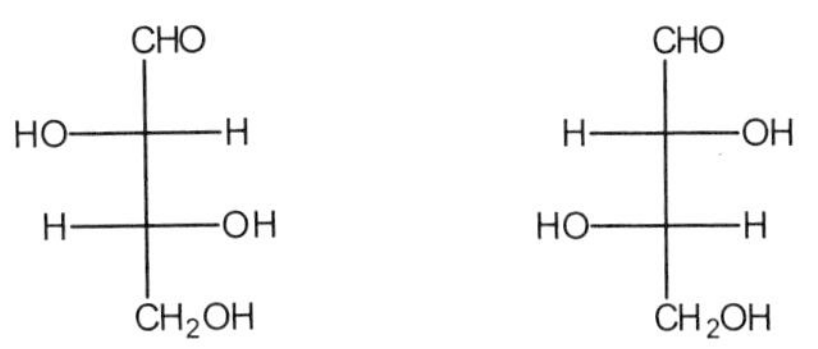

4. For chemical reactions leading to a racemic mixture, it is common to draw only one out of two possible enantiomers. The note "plus an enantiomer" on the drawing is nice, but should not be always expected.

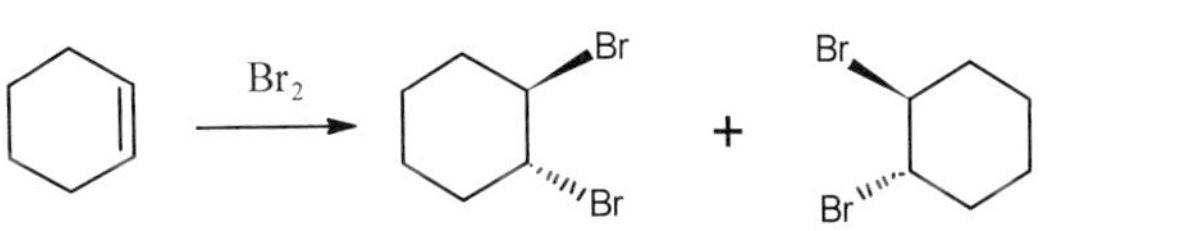

Pushing the limits

The concept of chirality is not limited to a chiral center. The example below shows two enantiomers with no chiral centers.

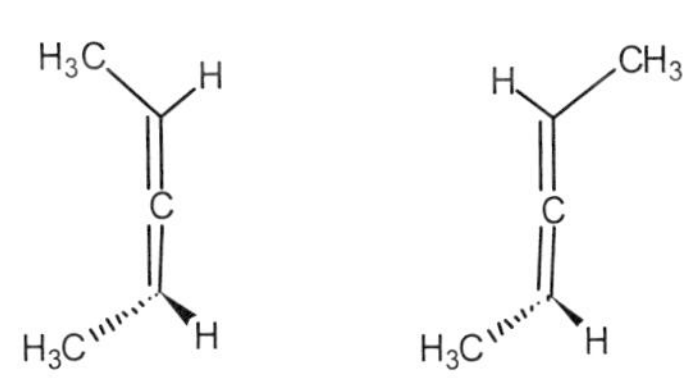

Chirality is introduced to these compounds by the *chiral axis* running through both C=C bonds.

The plane of symmetry is not the only element of symmetry that makes an object achiral. This concept is discussed in detail in any crystallography course.

If an atom of carbon is bonded to three different groups, and the fourth group is a mirror image of one of the other three groups, then those four groups are all different, but not by much. As a consequence, the concept of the chiral center does not work. Such situations are discussed in advanced stereochemistry courses in terms of *pseudocharality*.

Practice problems

1. **Which of the following compounds is the enantiomer of compound A?**

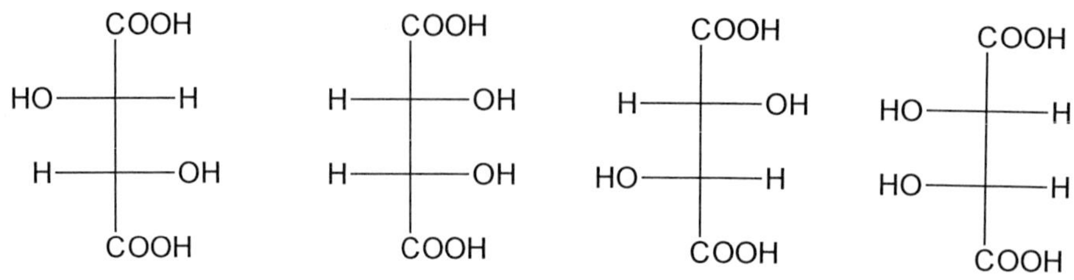

A B C D

2. The enantiomers from question 1 reacted with excess of (2R)-butan-2-amine. **What type of salts was produced?**

A. Enantiomers **B.** Diastereomers **C.** Constitutional (structural isomers)
D. Conformers **E.** Carboxylic acids

3. The enantiomers from question 1 reacted with excess of butan-1-amine. **What type of salts was produced?**

A. Enantiomers **B.** Diastereomers **C.** Constitutional (structural isomers)
D. Conformers **E.** Carboxylic acids

4. The enantiomers from question 1 reacted with excess of (2S)-butan-2-amine. **What type of salts was produced?**

A. Enantiomers **B.** Diastereomers **C.** Constitutional (structural isomers)
D. Conformers **E.** Carboxylic acids

5. Which of the structures from question 1 refer to the same compound?

A. A and B **B.** A and C **C.** B and D **D.** B and C **E.** C and D

6. Orient the molecule of compound I, so that Br is on the top and is pointing away from you, and the COOH group is on the bottom and pointing away from you. **Circle on the Fischer projection that represents <u>the orientation</u> of that molecule you just built?**

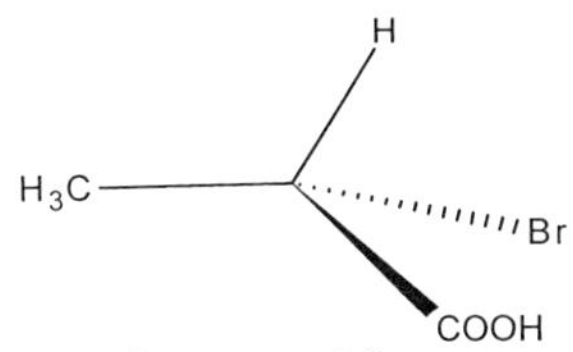

Compound I

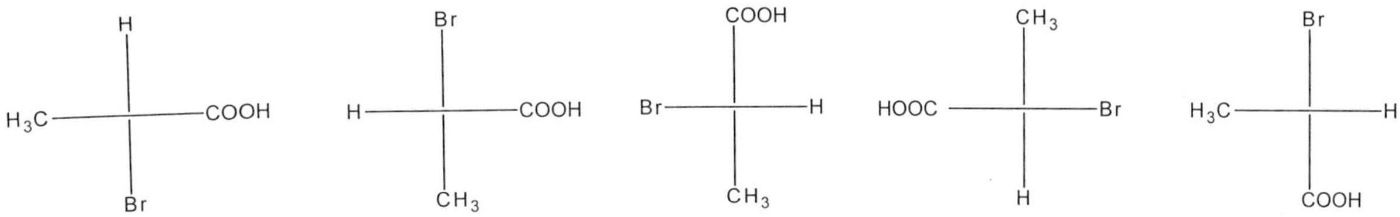

7. **Assign configuration to the carbon, which will lose Br after the following reaction**

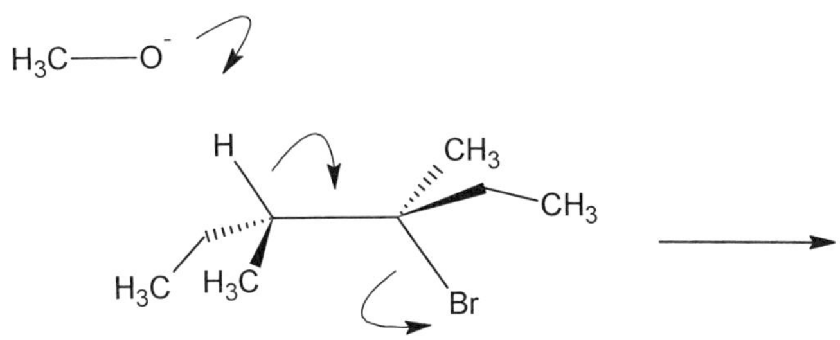

A. S- **B.** E- **C.** syn- **D.** anti- **E.** R-

8. **Circle the chiral center that loses its chirality in the following reaction**

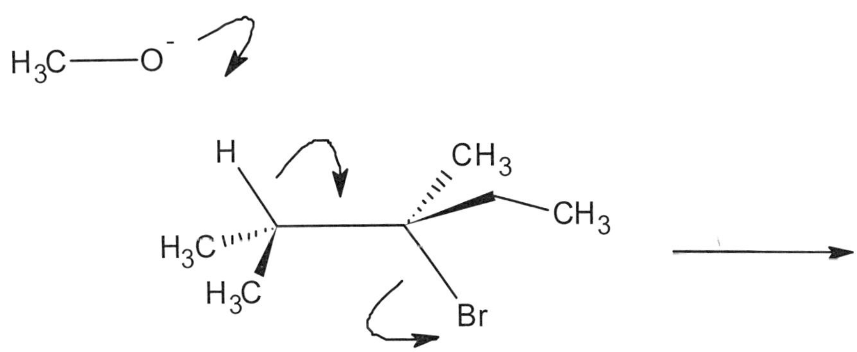

9. **"The following structures are diastereomers". What is your opinion about this statement?**

A. It is not correct, because they are mirror images

B. It is correct, because the chiral center has same configuration in these structures

C. It is correct, because the chiral center has opposite configuration in these structures

D. It is not correct, because the C=C bond has the R-configuration in these structures

E. It is correct, because the C=C bond has the Z-configuration in these structures

10. Circle the highest priority substituent in the following compound:

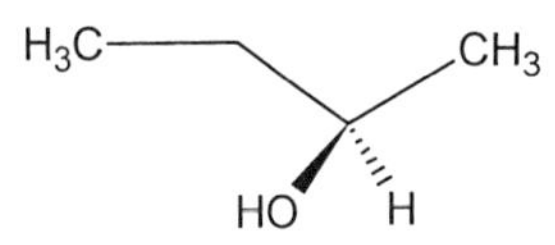

Making connections

First, all reactions from the previous chapters should be revisited from the stereochemical perspective, which also provides a better insight into the reactions covered later in this book. This chapter provides an indispensable foundation for learning the chemistry of aminoacids and carbohydrates.

Chapter 5

Aliphatic Halides, Alcohols, Ethers and Amines

This chapter is based on your knowledge of stereochemistry, and will introduce you to a set of reactions of substitution and elimination. First, we will prevent a few confusions about the terminology.

"Aliphatic" means "derivative of an alkane". However, the chemistry of aliphatic compounds, and sometimes the term "aliphatic", also applies to compounds with non-aliphatic fragments separated from the reaction center by at least three carbons. For example, 1-bromo-3-phenylpropane usually behaves like an aliphatic bromide.

If a compound has a OH-group attached to an sp^3-carbon, it is an alcohol. Compounds with an OH-group attached to an sp^2-carbon of a carbon-carbon double bond, or an sp-carbon of a carbon-carbon triple bond are vinyl alcohols which are usually unstable, or alkynyl alcohols. However, when an OH-group is attach to an aromatic ring, chemical properties of the compound are so different from those of an alcohol, that the compound is considered as a *phenol* rather than an alcohol.

Ethers are produced by dehydration of alcohols, phenols, or both, when a molecule of water is formed by a H abstracted from one OH-group, and OH of another OH-group. The two freed valences form a chemical bond. Both participating OH-groups may belong to either different molecules, or to the same molecule. The latter case is a cyclic ether.

The assignment of a compound to a 1°, 2°, or 3° type can be tricky, because *for amines, we count the number of carbons connected to nitrogen, while for all other compounds, we count carbons connected to the carbon bearing the functional group.* For example, tert.butyl amine is 1°, tert.butyl alcohol is 3°, tert.butyl chloride is 3°, and dimethylamine is 2°.

Solvents and hydrogen bonds

The overwhelming majority of chemical reactions are carried out in a solvent. Interaction *(solvation)* of a *solvent* with a *solute* (a dissolved compound) results in an associate *(solvate)*, which is the actual reactant in a reaction in *solution*. Solvation affects chemical reactions in a solution mainly by the following ways:

1. Solvation of starting compounds drives their solubility, and makes the solution reaction possible.

2. Solvation of transition states and intermediates decreases their energies and, therefore, increases the reaction rates.

3. Poor solvation of reaction products leads to their precipitation, which shifts the equilibria toward the products.

It should not be overlooked that solvation requires disconnection between different molecules of the solvent and disconnection between different molecules of the solute and, therefore, works against energies of those interactions.

More polar solvents tend to dissolve more polar compounds. Low polar solvents do not dissolve well highly polar compounds, because the energy of solvation does not make up for disconnecting tightly interacting polar molecule of the solute. Highly polar solvents do not dissolve well low polar compounds, because the energy of solvation does not make up for disconnecting tightly interacting polar molecule of the solvent.

When Bronsted acids interact with bases, they do not always donate protons. Rather the protons shift electrons from the base towards themselves, forming a weak bond. This bond is called a *hydrogen bond*, and is typical for hydrogens connected to a highly electronegative atom, such as O, N, or halogen. The strongest hydrogen bonds are formed with Lewis bases donating electrons from an element of the second period of the Periodic Table, which is, practically, O, N, or F. The Lewis acid component of a hydrogen bond is termed *hydrogen bond donor*. The Lewis base component is termed *hydrogen bond acceptor*.

The liquid hydrogen bond donors are termed *protic solvents*. The most common protic solvents are water, alcohols, primary and secondary amines. Protic solvents are always polar: the partial positive charge is localized on the hydrogen, and the partial negative charge is localized on the electronegative atom connected to that hydrogen. Due to the lack of steric hindrance, both the positive and negative parts of molecules of a protic solvent are easily accessible by the solute. Therefore, protic solvents efficiently solvate both cations, and anions.

Not all polar solvents are protic. Most common examples of polar aprotic solvents are acetone, dimethylformamide (DMF), and dimethylsulfoxide (DMSO).

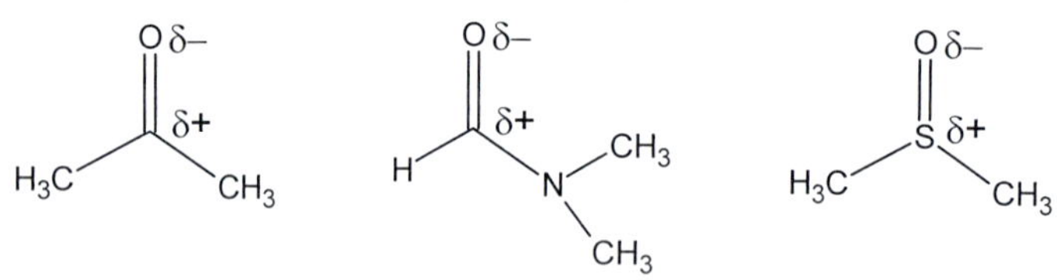

Acetone Dimethylformamide Dimethylsulfoxide

Due to the steric hindrance, the negative parts of molecules of an aprotic solvent are more accessible by the solute than the positive parts. Therefore, aprotic solvents efficiently solvate cations, and much less efficiently - anions.

The reaction mechanisms S_N1, S_N2, E1, and E2 are well explained in any organic chemistry textbook and summarized in the commented animations:

http://ochem.orgfree.com/sn1.html

http://ochem.orgfree.com/sn2.html

http://ochem.orgfree.com/E1.html

http://ochem.orgfree.com/E2.html

Before delving into the mechanistic details, it is important to keep in mind a few general considerations that will keep you from unnecessary memorization.

First the reaction mechanisms that start from "S" are Substitutions, and the mechanisms that start from "E" are Eliminations.

Second, any elimination always frees some valences, which connect together and form either a multiple bond, or a cycle.

Third, in any substitution, two events always happen by definition: 1. A part of the molecules cleaves off (a leaving group). 2. Another group enters the molecule. A specific substitution mechanism is nothing more than a graphical explanation of how exactly those two events are happening. The adjective at the reaction mechanism characterizes the entering group. When this group is a nucleophile, a reaction is Nucleophilic substitution, and the letter "N" shows up in the symbol of the mechanism. Three possible scenarios of substitution are logically possible:

1. *The entering group enters the molecule first, and the leaving group leaves the molecule next.* This mechanism is impossible at sp^3-carbons, which are already connected to the maximum number of atoms (4), so cannot accept any entering groups.

4. *The leaving group leaves the molecule first, and the entering group enters the molecule next.* This mechanism is possible at sp^3-carbons, freeing up a valency on the first step, which is next filled in with the entering group. This description explicitly suggests two reaction steps in the reaction mechanism. S_N1 is a mechanism of this type.

4. The entering group enters the molecule, and the leaving group leaves it about simultaneously, in one reaction step. S_N2 is a mechanism of this type.

The number in the reaction mechanism abbreviation must not be confused with the number of reaction steps. Rather, it stands for the reaction kinetic order. If concentrations of all reactants are increased by the factor of C, the reaction rate increases by the factor of C^n. The number *n is the reaction's kinetic order.* The most apparent practical consequence is that if reactions of different kinetic orders are competing, concentrating the solution favors the reaction of a higher kinetic order, while diluting the solution favors the reaction of a lower kinetic order.

The value of the kinetic order is usually close to the number of reacting particles in the significantly slowest (rate limiting) step. The S_N1 and E1 reactions share the rate limiting step of the formation of a carbocation from one molecule of the reactant. For both S_N2 and E2 reactions, the molecule of a reactant must collide with a particle of a nucleophile or a base. Therefore, the kinetic order is close to 2 for both reactions.

What else besides of concentrations of the reactants on the rate limiting step affects the overall reaction rate? Anything involved in the rate limiting step: intermediates, bases, nucleophiles, solvents. Now, with this knowledge in hand, you can look up the list of factors affecting S_N1, S_N2, E1, and E2 reactions to discover that you do not need to memorize them. You understand them now.

The nucleophilic reactions are based on the concept of nucleophiles and nucleophilicity. Just like bases, nucleophiles donate a pair of electrons. Donation of a pair of electrons to a hydrogen atom is usually well described in terms of basicity. However, the process of donating an electron pair to a non-hydrogen atom is so significantly affected by the electron pair acceptor, that the concept of basicity does not work. The most common remediation of this problem in organic chemistry is the *concept of the nucleophile*. The strength of a base is viewed as its ability to *shift the equilibrium* toward its conjugate acid, and is well correlated with the pK_a value of its conjugate acid. Nucleophilic reactions are much more diverse than acid-base reactions, so there is no single characteristic of the nucleophile to characterize its effectiveness. Nevertheless, the concept of the *strength of a nucleophile (nucleophilicity)* does exist, but applies only to a specific nucleophilic reaction – S_N2 reaction. *A stronger nucleophile makes the S_N2 reaction* *proceed faster*. Please, note that the concept of basicity is based on thermodynamics, while the concept of nucleophilicity is based on kinetics.

It is important to understand that not all nucleophilic reactions are affected by the nucleophilicity if the nucleophile. Nucleophilicity of the nucleophile is critical for the S_N2 reaction, and beneficial for the S_N1 reaction. However, the acyl nucleophilic substitution and aromatic nucleophilic substitution are promoted by the *basicity of the nucleophile*. Nucleophilic addition can be sensitive to both basicity and nucleophilicity of the nucleophile. There is a type of aromatic nucleophilic substitution, which even relies on the redox potential of the nucleophile.

Now, let us focus on the S_N2 reaction to figure out what affects nucleophilicity. First, a nucleophile large in size has a more diffuse and mobile electron clouds that can be easily deformed (polarized) by the electron pair acceptor. As a result, the ability of the nucleophile to donate electrons significantly increases in the process of the S_N2 reaction. In this sense, large nucleophiles behave as "cheating bases". That is why extremely weak bases, such as I^-, can be excellent nucleophiles. This effect of "hidden nucleophilicity", however, does not override the ability to donate an electron pair as determined by basicity, so a stronger base would still be a stronger nucleophile. Please, *do not memorize this trend*, because it is true only when the nucleophile is not affected by the interaction with the solvent (solvation), which happens either in vacuum, or in a solvent that does not significantly solvate anions (polar aprotic solvents such as acetone or dimethylsulfoxide). Most common solvents are polar (water or ethanol), which interact with nucleophiles by accepting their electron pair. It means, that the ability to donate an electron pair as a base is mostly "drained" by solvation. The "hidden" nucleophilicity of large nucleophile becomes, therefore, the predominant factor. So, *in protic solvents, nucleophilicity increases with the size of the nucleophile, or from the top to the bottom in the groups of the Periodic Table.*

When we look at the elements of the same period of the Periodic Table, their sizes are similar, so a stronger base is a stronger nucleophile. Both basicity and nucleophilicity increase from the right to the left in the periods of the Periodic Table (trend opposite to the trend of electronegativity), and with the increase of the negative charge (which obviously helps to donate negatively charged electrons).

<u>Trend examples for protic solvents</u>:

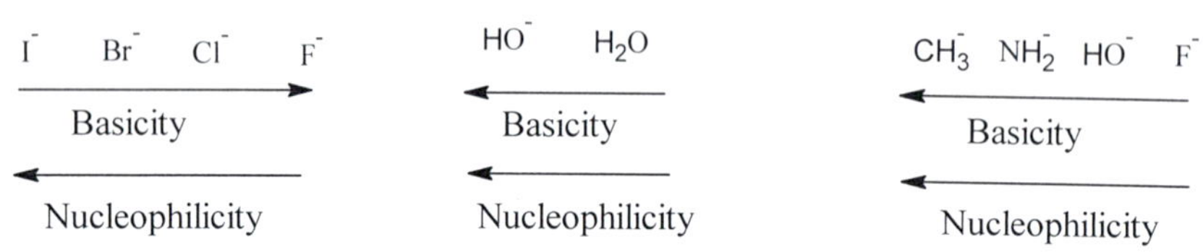

Since the rate limiting step of the S_N1 reaction produces both cations and anions, the *protic solvents able to solvate both, are better for carrying out S_N1 reactions.*

Since the S_N2 reaction often involves nucleophilic anions, the *aprotic solvents not deactivating them by solvation, are better for carrying out S_N2 reactions.*

At this point, you are referred to any organic chemistry textbook for a detailed coverage of this topic. After you are done, let us look at a few practical applications of your knowledge in organic synthesis.

1. Williamson synthesis of ethers

By the Williamson's method, ethers are synthesized through the S_N2 reaction with an alkoxide as the nucleophile:

$R_1\text{-}O^- + R_2\text{-}I \longrightarrow R_1\text{-}O\text{-}R_2 + I^-$

To choose whether the R_1 or R_2 part of the target ether will originate from the alkoxide, we must consider competition with elimination. Do not choose a tertiary R_2. It is too hindered for the S_N2 mechanism, and so reactive in the E2 mechanism (the incipient C=C bond in the transition state is stabilized by at least two alkyl substituents), that the S_N1 reaction is too slow to compete. If the R_2 group is CH_3 or primary, the Williamson synthesis has a good chance to succeed.

2. Tuning the non-rate-limiting step

If two competing reactions are sharing the rate limiting step, we cannot favor either one by affecting that step. In those cases, to change the reaction outcome, we must influence a less sensitive, non-rate-limiting step. For instance, the S_N1 and E1 reactions are sharing the rate limiting step.

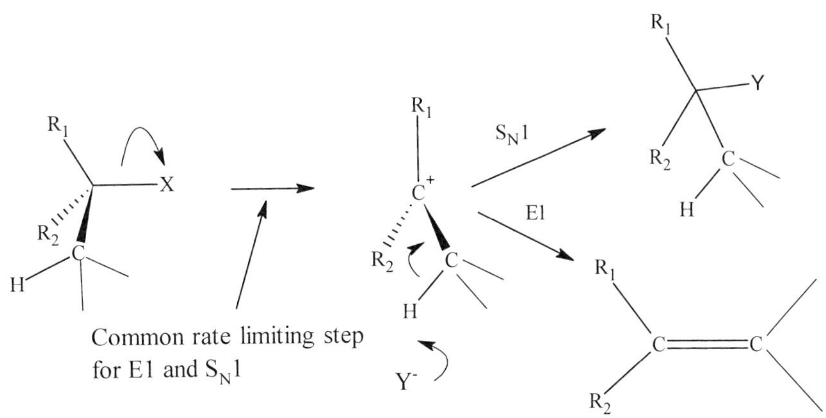

The following example shows how we can shift the reaction outcome to the S_N1 reaction by using a stronger nucleophile (Cl^-) at cooling, while under heating in the presence of a weaker nucleophile (SO_4^{2-}), the reaction results in the E1 product:

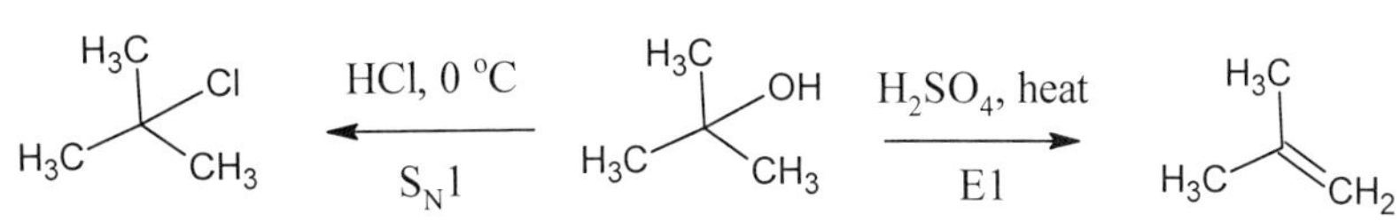

Elevated temperature usually favors elimination because of its higher enthropy and the lower boiling point of the alkene that escapes from the reaction mixture.

3. Activating the leaving group

The leaving group is involved in the rate limiting step of all four reactions in question. In order to run them with an alcohol, the OH^- group (strong base, bad leaving group) is usually converted to a good leaving group (H_2O, weak base, good leaving group) by protonation. For example:

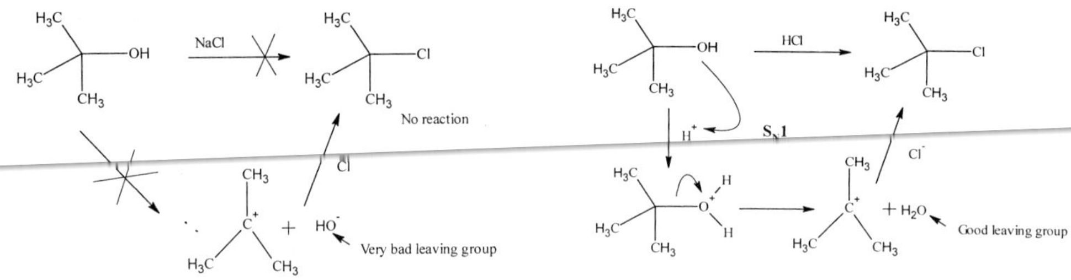

Same approach applies for all three remaining reactions.

4. Activating the leaving group and choosing the solvent

Often, chemists need to substitute the hydroxy-group at a chiral center and avoid racemization. Therefore, the reaction condition must be designed to activate the hydroxy-group and favor the S_N2 pathway over the S_N1 pathway. However, polar aprotic solvents that work best for S_N2 reactions, usually react with acids. Therefore, we cannot use them to activate alcohols by protonation. In this situation, we can activate the alcohol by reacting with an acid chloride of a very strong acid to form its ester. This reaction does not affect the C-O bond and, therefore, keeps the chiral center in place. The leaving group is now very good, because it is the conjugate base of a very strong acid. Next, we perform the S_N2 reaction with the appropriate nucleophile in a polar aprotic solvent.

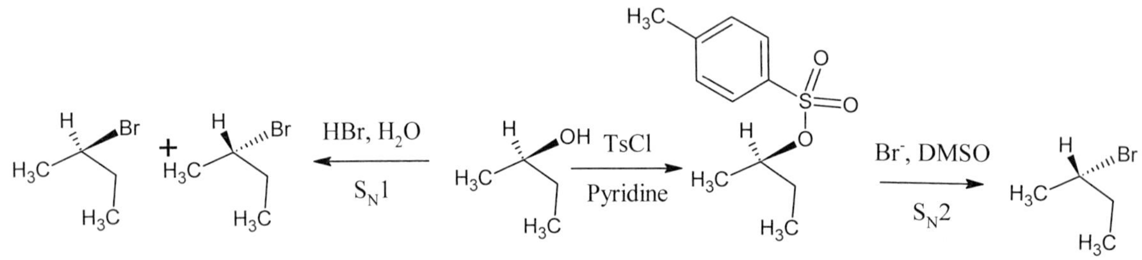

This alternative method activates alcohols without protonation, which is also very useful for acid-sensitive alcohols.

Here is a "radically" different reaction for preparation of alkyl *halides: free radical halogenation.*

The chain reaction mechanism (http://ochem.orgfree.com/SR.html) consists of three groups of steps:

a. Initiation

This group of steps *produced free radicals (particles with an unpaired electron) from non-radicals.* Most commonly, the free radicals are generated by splitting a halogen molecule by light (as shown in the animation), or splitting a weak O-O bond in peroxides by heat.

b. Propagation

This group of steps *converts a free radical to another free radical.* For the free radical halogenation, the halogen radical produced on the second propagation step, participates in the first propagation state with the next molecule of the hydrocarbon:

$RH + Br\cdot = R\cdot + HBr$; $R\cdot + Br_2 = RBr + Br\cdot$

c. Termination

When the first propagation step runs out of the starting hydrocarbon, or even before that, two *free radicals form a covalent bond and turn into a non-radical particle*. This is the termination group of steps. When the overall rate of termination exceeds the overall rate of initiation, the chain reaction slows down and stops, even before the starting reactants have run out.

What can go wrong?

At this point, the missing skill to relate the general reaction mechanism to a specific example usually returns with revenge. Here is how to practice it:

1. Write a reaction mechanism with curved arrows. Flip structures of the starting materials by 90°, and write the same mechanism from the right to the left. Change the structure orientation again, and keep practicing until you are confident to write the reaction in any direction regardless of how the structures are drawn.

2. Find (Google for example) a structural formula of a prescription medication, which is likely to have a fairly complicated structure. Identify in that formula a functional group, for which you can write a reaction mechanism. Draw that mechanism, simply carrying over the rest of the structure through all steps. Find other functional groups, and decide which of them also can be affected by the same reaction conditions. Write those reaction mechanisms.

Now, think of the following statement: "By looking at the abbreviation E2, it is clear that that reaction proceeds in 2 steps". I you think that the statement is correct, you need to read this chapter again. Otherwise, proceed to the next paragraph.

Critical thinking is essential here. When you are assessing the reaction rate by the stability of a carbocation, do not overlook stability of its counter-anion. When you are assessing acidy by the stability of the conjugate base, do not overlook the strength of the bond between the atom of hydrogen and the rest of the acid molecule. A good instructor will never be offended if you ask about alternative considerations contributing to the whole picture.

For the chain reactions, think of a single particle when you are writing the propagation steps. For instance, for free radical bromination of alkanes, students often think that all alkane is consumed on the first propagation step. It makes unclear why the Br radical released after the second propagation step react with the alkane again. This potential confusion can be averted by the understanding that after one molecule of alkane goes through the propagation step, the remaining molecules are available to continue the chain reaction.

Pushing the limits

For the next step of learning organic chemistry, any reaction mechanisms such as S_N1, S_N2, and E2, should be viewed as models rather than the set in stone options. For instance, there is an infinite number of substitution mechanisms between S_N1 and S_N2. It is important to assess how well a specific process in described by the chosen model or models.

Practice problems

1. The following compounds rank in the order of their decreasing reactivity toward CN^- as follows: B > C > A > D > E.

   A. 1-Chloropropane **B.** 1-Iodopropane **C.** 1-Bromopropane **D.** 1-Fluoropropane **E.** 1-Propanol

Acid	pK_a
HI	-9
HBr	15.7
HF	-7
H_2O	3.2
HCl	-11

2. **Rank the following compounds in the order of their decreasing reactivity toward H_2O (The most reactive compound is listed first).** Use your answer to question 1.

   A. 2-Chloro-2-methylpropane **B.** 2-Bromo-2-methylpropane **C.** 2-Fluoro-2-methylpropane

   D. 2-Iodo-2-methylpropane **E.** 2-Methyl-2-propanol

3. What were you comparing to answer questions 1 and 2?

   A. Nucleophilicity of nucleophiles

   B. Stability of carbocations

   C. "Quality" of the leaving group (its ability to leave)

   D. Steric hindrance

   E. Stability of the reaction product

4. The following compounds rank in the order of their decreasing reactivity toward in S_N1 reactions (after protonation) as follows:

   C > D ~ E > A > B. ("~" Means "Reactivity is close (similar)).

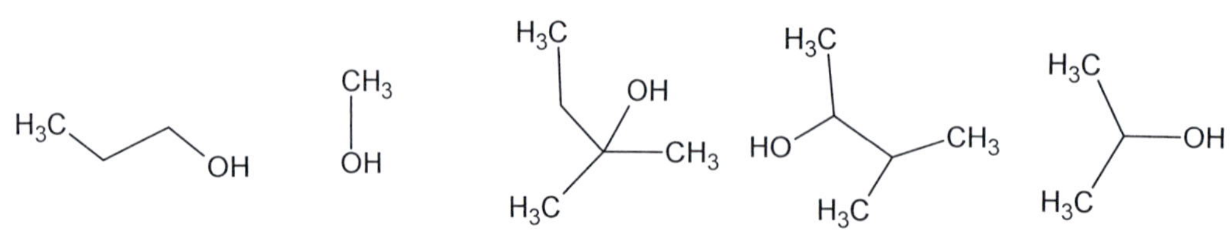

A B C D E

What was compared in order to deduce the ranking above?

   A. Nucleophilicity of nucleophiles

   B. Stability of carbocations

   C. "Quality" of the leaving group (its ability to leave)

   D. Steric hindrance

   E. Stability of the reaction product

5. Which of the following concepts is illustrated by your answer to question 4? (All the following statements are correct)

   A. The E2 reaction usually proceeds in the anti-fashion

   B. The S_N1 reaction proceeds through a carbocation on its rate-limiting step

C. The nucleophile is involved into the rate-limiting step of the S_N2 reaction

D. The S_N1 reaction proceeds in two steps

E. The E2 reaction is highly preferred over the S_N1 reaction for tertiary halides in the presence of a strong base

6. Rank the following nucleophiles in the order of their decreasing reactivity in S_N2 reactions (The most reactive nucleophile is listed first).

A. Cl^- **B.** H_2O **C.** F^- **D.** I^- **E.** Br^-

7. Which of the following concepts is illustrated by your answer to question 6? (All the following statements are correct)

A. The leaving group is involved into the rate-limiting step of both S_N1 and S_N2 reactions

B. The S_N1 reaction proceeds through a carbocation

C. The nucleophile is involved into the rate-limiting step of the S_N2 reaction

D. The S_N1 reaction proceeds in two steps

E. The E2 reaction is highly preferred over the S_N1 reaction for tertiary halides in the presence of a strong base

8. Light-induced reaction of alkanes with Br_2 proceeds through free radicals and produces alkyl halides. **Circle the bond that is cleaved <u>by light</u>.**

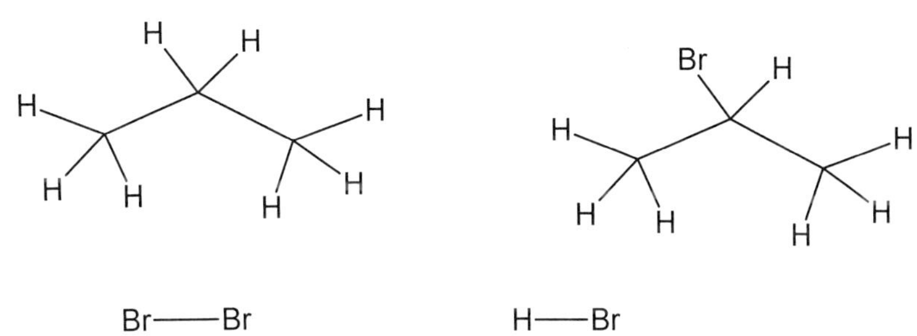

9. Alcohols are stronger acids, than alkanes by about 40 orders of magnitude. **Circle the atom, whose high electronegativity explains why it happens.**

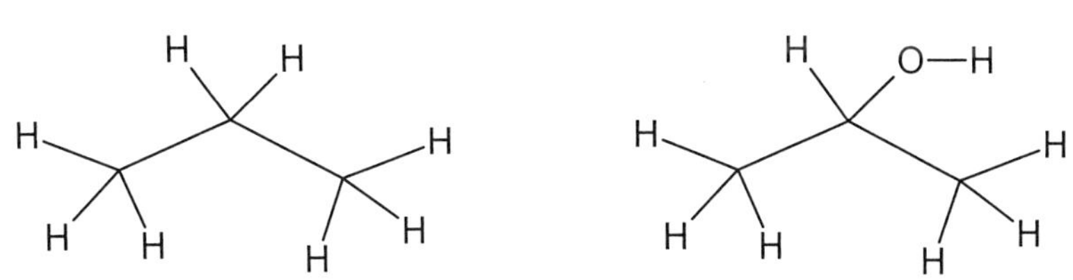

10. For tertiary substrates, the E2 reaction is overwhelmingly favored over S_N1 in the presence of a strong base. **Circle the compound that will produce 2-methylpropene upon treatment with sodium methoxide.**

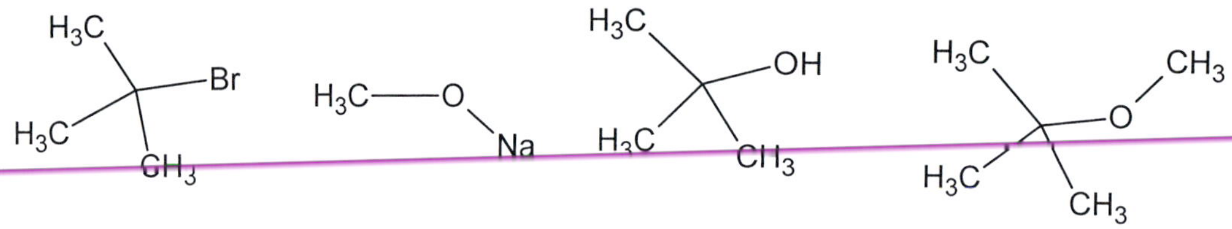

11. How would you attempt to remove an impurity of a water-insoluble thiol from a water-insoluble alcohol?

A. Washing with water **B.** Washing with aqueous HCl

C. Washing with aqueous NaOH **D.** Adding water and then filtration

E. I will not use relative acidity of alcohols and thiols to solve this problem

Making connections

Knowledge of the S_N1, S_N2 reactions will lay groundwork for learning nucleophilic addition to the carbonyl group and substitution in aromatic compounds. From the practical perspective, this chapter will help understand synthesis of alkynes from acetylene. The prior knowledge of stereochemistry will help with modification of chiral alcohols, while the orbital theory and acidity concept have helped you understand hydrogen bonding and chemistry of alkoxides. Oxidation of alcohols links the topic to the chemistry of carbonyl compounds that will be expanded later in the book.

Chapter 6

Conjugation and aromatic compounds

Once you learnt chemistry of compounds with one multiple carbon-carbon bond, and gained experience with the reactions of addition, substitution, and elimination, we are now ready to place several carbon-carbon bonds in one molecule, and see how it may spice up its chemistry.

Two C=C bounds next to one another in the molecule of 1,3-pentadiene are connected with a single carbon-carbon bond, so the π-components of the C=C bonds can align by rotation around the single bond. As a consequence, additional bonding between carbons stabilizes this conformation and, therefore, the whole molecule. In this type of dienes (1,3-dienes), the C=C double bond between carbons 1,2 and 3,4 significantly interact (by π-conjugation), which is why they are also referred to as *conjugated dienes*. Higher stability of 1,3-dienes is demonstrated by their lower heats of combustion comparing with isolated dienes.

Heats of combustion:

CH2=C=CH-CH2-CH3	$CH_2=CH-CH_2-CH=CH_2$	CH2=CH-CH=CH-CH31,2-pen
1,2-pentadiene (1,2-diene)	1,4-pentadiene (isolated diene)	1,3-pentadiane (1,3-diene)
3251 kJ/mol	3217 kJ/mol	3186 kJ/mol
Least stable		Most stable

Weak σ–π conjugation* *Better σ–π* conjugation* *Both σ–π* and σ–π* conjugations* **=C=** is a new example of sp-hybridized carbon

Contrary to 1,3-dienes, *in 1,2-dienes,* when one carbon is connected by two C=C bonds, *there is no conjugation* between the two p-components, because they are located in two perpendicular planes.

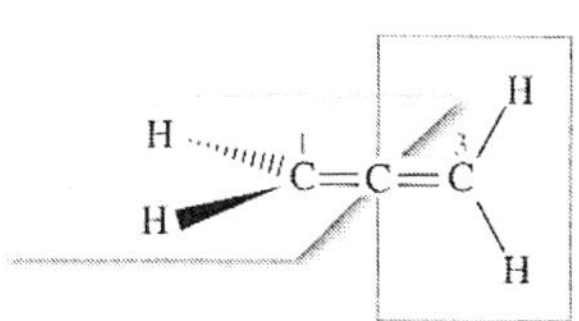

The reason of such geometry is that at an sp-carbon, two p-orbitals participating in the formation of two C=C double bonds, are perpendicular to one another.

Please, note that the numbers in the classification of dienes (start from the first carbon of a C=C bond) do not have to match the numbers of the IUPAC nomenclature (start from an end of the principal chain). For instance, 2,4-hexadiene belongs to the type of 1,3-dienes, and 2,3-hexadiene belongs to the type of 1,2-dienes.

Because the conjugation of C=C bonds in 1,3-dienes decreases the energy, in the local minima stable conformations, the π-components of those bonds are maximally aligned. These eclipsed conformations around the bond between carbons 2 and 3, corresponding to the dihedral angles of 0° and 180° are shown on in the figure below. The *s-trans*-conformation has a lower energy than the *s-cis*-conformation because of a lower steric hindrance.

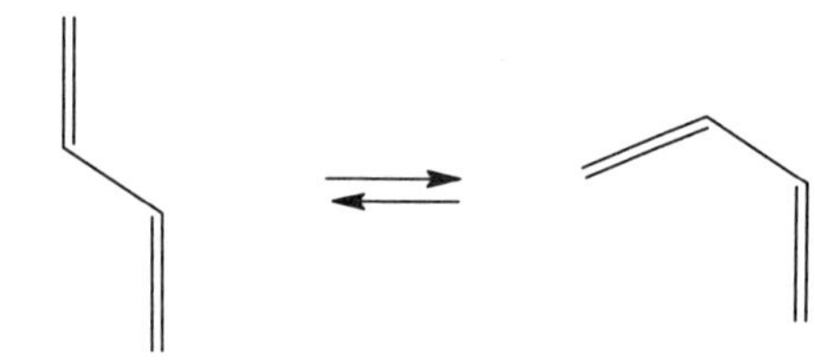

***s-trans*-conformation** ***s-cis*-conformation**

Perhaps, the most prominent feature of 1,3-dienes is 1,4-addition. *Eventually, carbons 1,4 end up to be bonded with new atoms, and a C=C bond appears between carbons 2 and 3.* This shortcut to derive the structure of the product from the structure of the 1,3-diene will work for all instances of 1,4-addition. Here are three most common examples:

1. Electrophilic hydrobromination

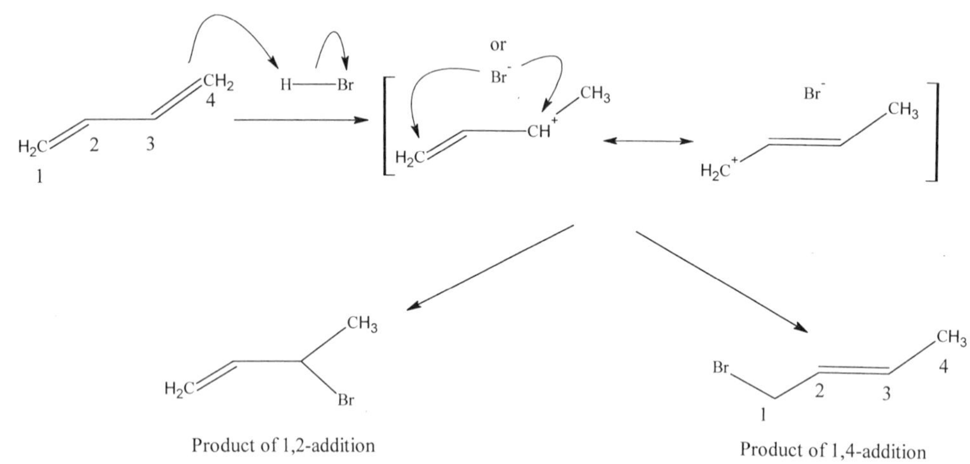

2. 1,4-Polymerization

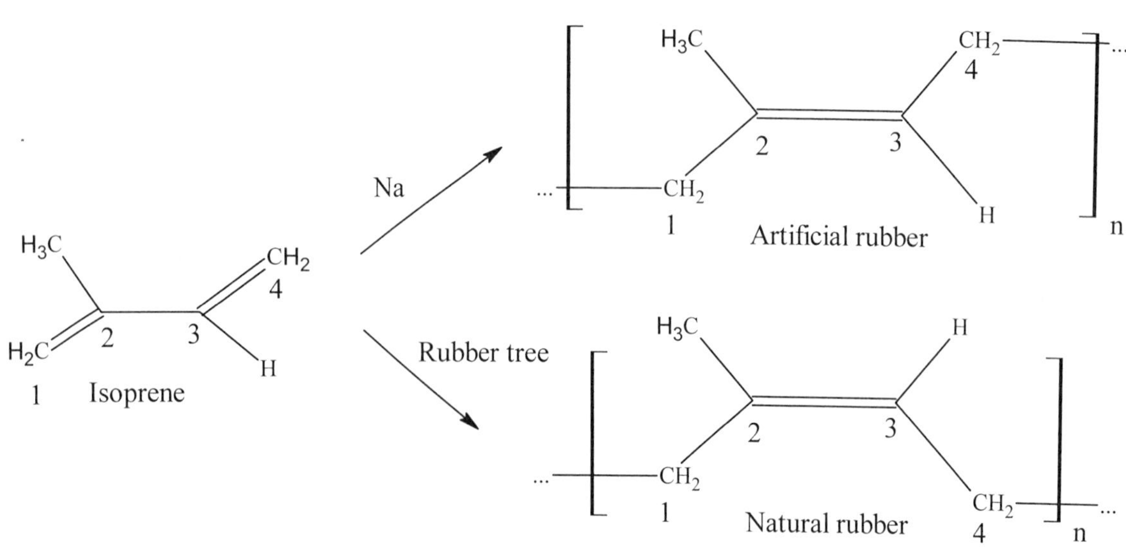

To correctly draw the structure of a 1,4-polymer of a 1,3-diene, based on the structure of the monomer:

1. Identify two carbons at the end of the C=C-C=C chain. They are carbons 1 and 4. The "inner" two carbons are carbons 2 and 3. Only those four carbons of the chain will become a part of the polymer chain.

2. If necessary, re-draw the structure of the monomer to remove all substituents from the horizontal line and avoid the common mistake of including those substituents to the polymer chain.

3. Break the π-components of both C=C bonds.

4. The freed valences of carbons 1 and 4 will hold together the monomer units in the polymer chain.

5. Draw the C=C bond between carbons 2 and 3.

6. Show the monomer units by the brackets, and extend the freed valances of carbons 1 and 4 over the brackets.

To correctly draw the structure of the monomer, based on the structure of a polymer:

1. Remove the brackets and identify four carbons of the polymer chain.

2. Connect atoms 1,2 and 3,4 with C=C bonds.

Please, note, that depending on the mechanism of addition, the new C=C double bond can have either *cis-* or *trans-*configuration.

3. *Diels-Alder addition*

Diels-Alder addition (http://ochem.orgfree.com/diels.html) is usually a synchronous 1,4-cycloaddition of an alkene (in this context it is referred to as "dienophile") to a 1,3-diene, forming a 6-membered ring. Due to the enormous synthetic power of this reaction, Diels and Alder in 1950 have been awarded with a Nobel Prize in chemistry.

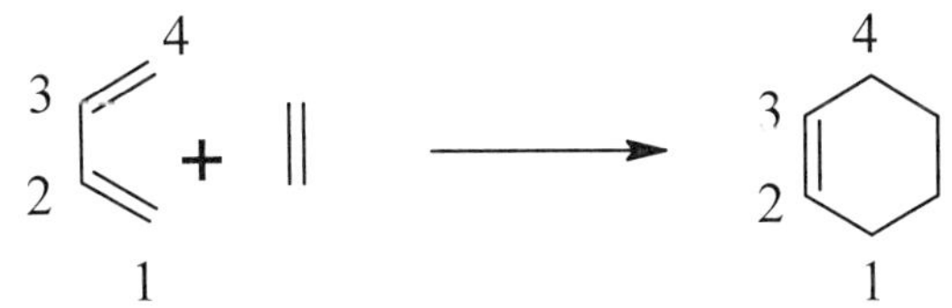

Please, note, that because of the angular strain, only the *s-cis*-conformation of a 1,3-diene actually undergoes the Diels-Alder addition, which results in the *cis*-configuration of the new C=C double bond.

Conjugated polyenes

Conjugated polyenes contain a chain of C=C double bonds arranged in the same way as in 1,3-dienes. The π-component of a single C=C double bond can absorb UV-light and the molecule turns into its excited state. Accumulation of C=C double bonds conjugated in a chain leads to stronger absorption of light, and the absorption band shifts towards lower energies, and at five C=C double bonds in conjugation, it crosses into the realm of visible light. The compound thus becomes colored, and the color is determined by which part of the visible light range is absorbed. All organic dyes have at least one chain of conjugated C=C double bonds.

Amazingly, interaction of conjugated C=C double bonds with light not only accounts for a significant part of the variety of colors, but also enables us to see those colors, or any visible light to that matter. Thus, visible light flips a molecule of cis-retinal, attached to a protein component (opsin), to its trans-configuration, which significantly changes the molecule's shape, pushes the opsin component to a different conformation, which eventually sends a signal to the brain. *Trans*-retinal is then converted back to its *cis*-diastereomer by an enzyme, and is ready to sense the next quantum of light.

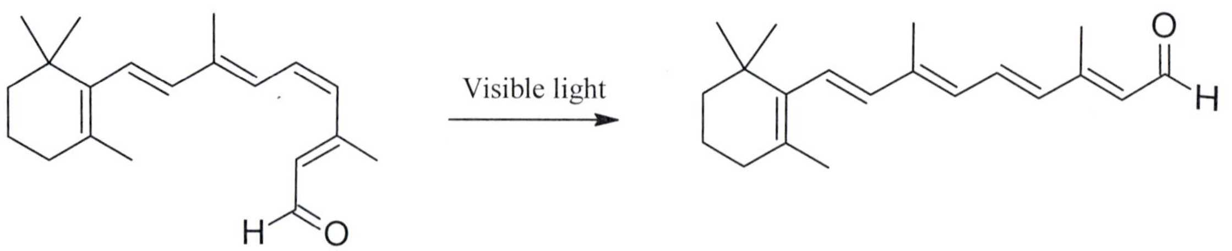

cis-Retinal *trans*-Retinal

Aromaticity

When we bring together the ends of a conjugated π-system, it may significantly affect its energy. A quantum chemistry computation would show that the system has an odd number of occupied π-orbitals, the cyclization leads to a significant drop in energy, which is referred to as *aromaticity*. Each of those odd (2n+1) orbitals holds 2 electrons, which bring the total number of electrons in cyclic conjugation to 2X(2n+1) = 4n+2, where n = 0,1,2,3,...etc. Therefore, **an aromatic system contains 4n+2 electrons in a cyclic conjugation.** This is *Huckel's rule.* Conversely, cyclization of a π-system with an even number of orbitals (2n) increases its energy, leading to an **anti-aromatic** case with 2X2n = **4n electrons** in a cyclic conjugation.

While the electron counting math is very easy, it is usually overlooked that all counted electrons must be in a **cyclic conjugation**. *First,* it means that there must be no sp^3-carbons in the aromatic cycle that would interrupt cyclic conjugation. *Second,* if an atom contributes a p-electron or an electron pair to the conjugation, it cannot contribute any other electrons, simply because the space is already occupied. *Third,* the best conjugation is achieved if the orbitals are aligned perpendicularly to the plane of the aromatic ring. Since aromaticity decreases energy, a molecule adopts a planar conformation if available. Since anti-aromaticity increases energy, a molecule avoids a planar conformation if possible.

Examples:

The lone electron pair on nitrogen in pyridine does not count for Huckel's rule, because, that nitrogen already contributes a p-electron to the conjugation.

Cyclopentadiene is non-aromatic, because the cyclic conjugation is interrupted by an sp^3-carbon.

Benzene and pyridine adopt the planar conformation and are aromatic. All six carbons in benzene are indistinguishable.

Cyclooctatetraene has a non-planar conformation. It escapes the anti-aromatic conformation and is non-aromatic.

Cyclobutadiene is not flexible enough to escape the almost planar conformation. It is anti-aromatic.

In pyrolle, the nitrogen with the lone electron pair adopts a planar geometry to maximize alignment of the lone electron pair with the rest of electrons in the cyclic conjugation. Pyrrole is aromatic.

In the cyclopentadienyl-anion, the carbon with the lone electron pair adopts a planar geometry to maximize alignment of the lone electron pair with the rest of electron in the cyclic conjugation. All five carbons in that anion are indistinguishable. This anion is aromatic.

In furan, one lone electron pair on oxygen counts for aromaticity, just like in pyrolle. The second electron pair cannot participate in conjugation, because the space required for the alignment, is already occupied with the first electron pair on oxygen. Furan is aromatic.

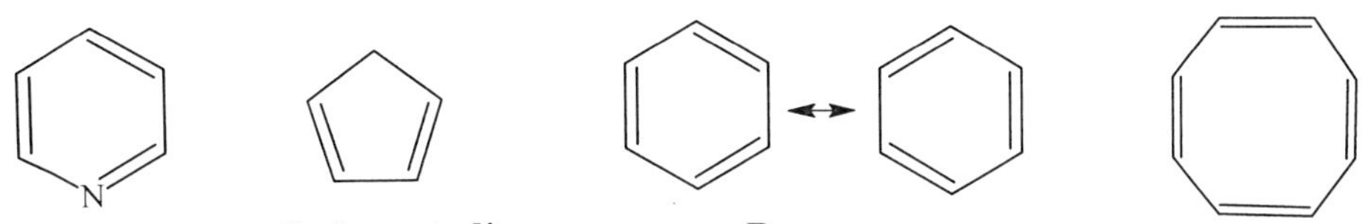

Pyridine, aromatic **Cyclopentadiene, non-aromatic** **Benzene, aromatic** **N=6**

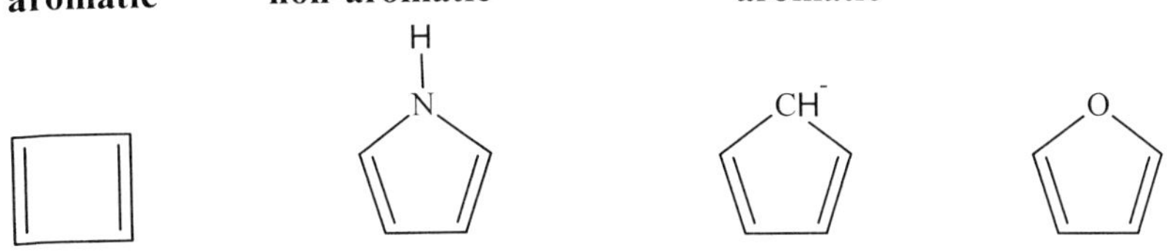

Cyclobutadiene, anti-aromatic **Pyrrole, aromatic** **Cyclpentadienylium-anion, aromatic** **Furan, aromatic**

The higher stability of aromatic compounds is illustrated by the following two reactions:

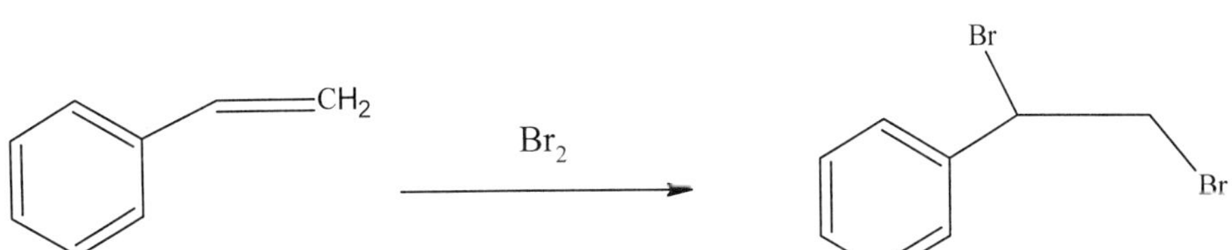

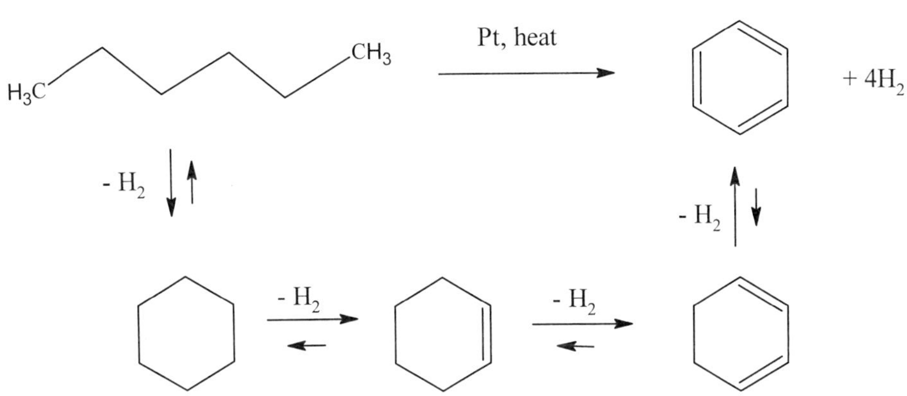

The first reaction shows that the benzene ring does not react with Br_2 under the conditions when a non-aromatic C=C double bond does. The second reaction shows that in a sequence of equilibrated processes, the overall equilibrium shifts toward a more stable aromatic compound. This is taken advantage of in a high-scale industrial process of making aromatic compounds from a straight-run gasoline during reforming.

Aromatic electrophilic substitution

The lack of reactivity of the benzene ring toward addition of Br_2 does not mean it would not react with Br_2 at all. Bromination of aromatic compounds requires a Lewis acid catalyst and proceeds as *electrophilic substitution as opposed to electrophilic addition.*

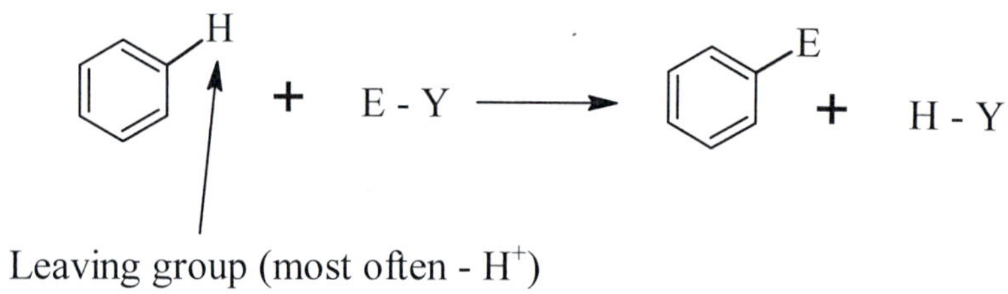

The mechanism of aromatic electrophilic bromination that proceeds through a non-aromatic σ-complex is presented on the following scheme, and is explained step-by-step in the animation: http://ochem.orgfree.com/SEAr.html

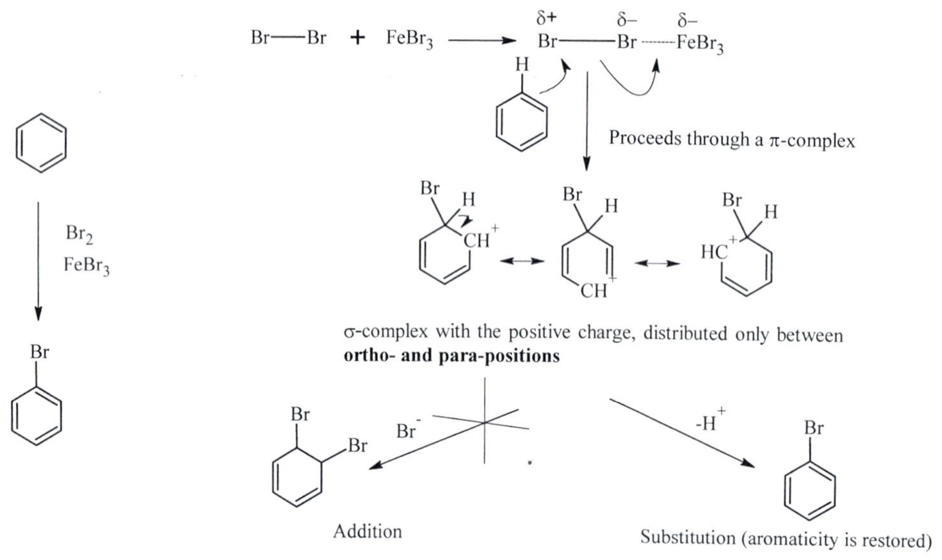

You can find many other examples of aromatic electrophilic substitution in any organic chemistry textbook. The centerpiece of understanding of any aromatic electrophilic substitution reaction is focusing on the non-aromatic σ-complex, which restores aromaticity on the next step of the reaction. Since the energy-favorable aromaticity is lost when the σ-complex is formed, this step is the rate-limiting, so a deeper insight in the structure of the σ-complex will let us assess both *reactivity* of compounds in this reaction, and identify the carbon, which is most likely to become bonded with the electrophile *(substituent directing effect).*

Looking at the s-complex is the key to understanding, why electrophilic *substitution* is characteristic for *aromatic* compounds, while electrophilic *addition* is typical for *alkenes*. Addition of a nucleophile to the σ-complex would create a second sp^3-hybridized carbon interrupting the cyclic conjugation in the ring, which makes the product of addition definitely non-aromatic. Contrary, deprotonation of the s-complex converts the only sp^3-carbon in the ring to an sp^2-carbon, and sends the electron pair from the breaking C-H bond to the conjugation. This restores aromaticity, and completely *shifts the reaction outcome from addition to substitution*.

When the compound that undergoes substitution already has one or more substituents, the remaining unsubstituted carbons are non-equivalent, so substitution at them leads to different products. The influence of substituent on which carbon receives the incoming electrophile is termed *"substituent directing effect"*. Both the substituent directing effect, and the substituent effect on the reaction rate are determined by the electronic effects of substituents, which are discussed in detail in other organic chemistry learning resources. Here, you will learn how to avoid the most common mistakes that students tend to make on this topic.

First, the terminology describing the positions of substituents in the ring with respect to one another sounds easy, but has a little pitfall.

Substituents at the carbons next to one another are said to be in the *ortho*-position.

Substituents at the farthest carbons of the ring are said to be in the *para*-position.

Substituents at the carbons separated by one carbon are said to be in the *meta*-position.

The *ortho-*, *meta-*, and *para*-positions must be determined with respect to another substituent, not to the top of the drawing of the molecule!

Second, there is no difference between the bonds in the benzene ring marked as "single" or "double", so the two drawings of ortho-dimethylbenzene on the following figure are identical.

ortho-Dimethylbenzene ortho-Dimethylbenzene meta-Dimethylbenzene para-Dimethylbenzene

It must be firmly understood, that the *existing substituent in the benzene ring, not the incoming substituent* during the substitution, mostly *determines the reaction rate and direction* of substitution. To understand why, we need to use our universal master key to most problems: look at the intermediate of the rate limiting step, which is the s-complex.

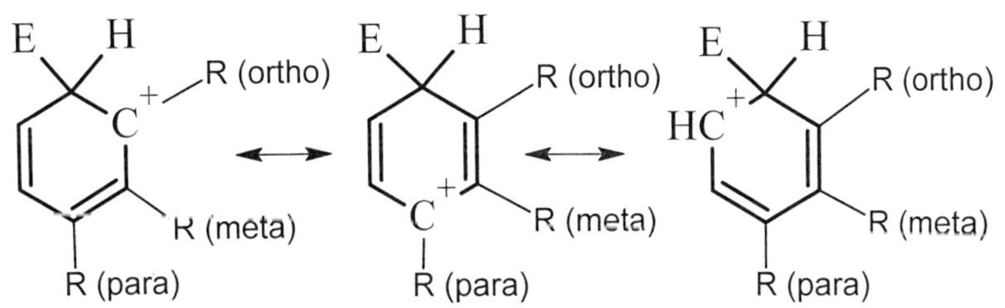

The *incoming substituent* (**E**) is bonded with an sp^3-carbon, which is not a part of the conjugated system, so it affects the reaction *the least*.

The *existing substituent* (**R (ortho), R (meta),** or **R (para)**) is bonded with a carbon participating in the conjugation, so it affects the reaction *much more.*

The *existing substituent* (**R (ortho)** or **R (para)**) bonded with a carbon participating in the delocalization of the positive charge (that is either *ortho-* or *para*-carbon with respect to the incoming substituent), affects the reaction the most.

Therefore, if the ***existing substituent is an electron donor,*** the σ-complex is stabilized, and the reaction proceeds faster ***(activating effect)***. Importantly, this activating effect is stronger when the existing substituent is in an ortho- or para-position with respect to the incoming substituent **(*ortho-para*-directing effect)**.

Conversely, if the ***existing substituent is an electron acceptor,*** the σ-complex is destabilized, and the reaction proceeds slower ***(deactivating effect)***. Importantly, this deactivating effect is stronger when the existing substituent is in an ortho- or para-position with respect to the incoming substituent, so those positions are deactivated more than the meta-position **(*meta*-directing effect)**.

However, usually a substituent has more than one effect, and things become a bit tricky.

For the *activating or deactivating effects*: Electron donor effects overpower electron acceptor effects, except for halogens where the electron acceptor effects are stronger. Examples: $-CH_3$ (hyperconjugative donor and inductive acceptor) – activator, $-OCH_3$ (resonance donor, inductive acceptor) – activator, -F (resonance donor, inductive acceptor) – deactivator, -Cl (resonance donor, resonance acceptor, inductive acceptor) – deactivator, $-NO_2$ (resonance acceptor, inductive acceptor) - deactivator.

For the *directing effects*: Electron donor effects always overpower electron acceptor effects. It means that if there is at least one donor effect, the substituent is ortho-para-director. Meta-director must be "pure" acceptors that is they cannot have any electron donor effects at all. Examples: $-CH_3$ (hyperconjugative donor and inductive acceptor) – *ortho-para*-director, $-OCH_3$ (resonance donor, inductive acceptor) – *ortho-para*-director, -F (resonance donor, inductive acceptor) – *ortho-para*-director, -Cl (resonance donor, resonance acceptor, inductive acceptor) – *ortho-para*-director, $-NO_2$ (resonance acceptor, inductive acceptor) - *meta*--director.

Allylic and Benzylic Reactivity

You are already familiar with a higher stability of allyl carbocation due to the conjugation of the orbital at the allylic position with the C=C double bond demonstrated by the resonance structures. Similarly, allyl free radical and allyl carboanions also have elevated stability, making the allylic position more reactive in general.

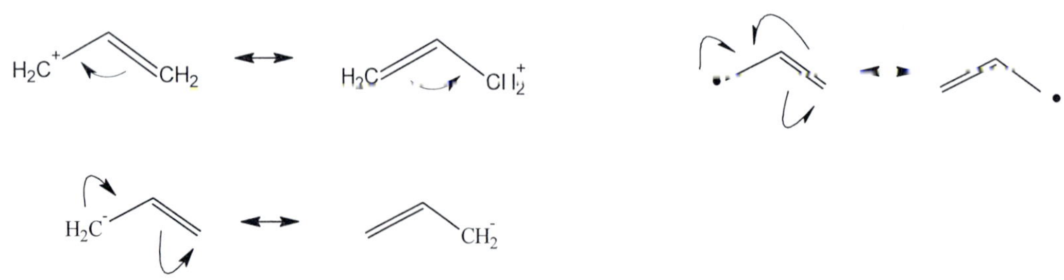

However, the term "*allylic*" is "carbon-centered": it applies differently to carbons and all other atoms, which can be tricky.

1. The **allylic carbon is a carbon bonded with an sp^2-carbon of a C=C bond.**

2. The **allylic group** or a non-carbon atom is a group or atom **bonded with an allylic carbon**.

The allylic position is not to be confused with the vinylic position.

1. The **vinylic carbon is an sp^2-carbon of a C=C bond.**

2. The **vinylic group** or a non-carbon atom is a group or atom **bonded with vinylic carbon**.

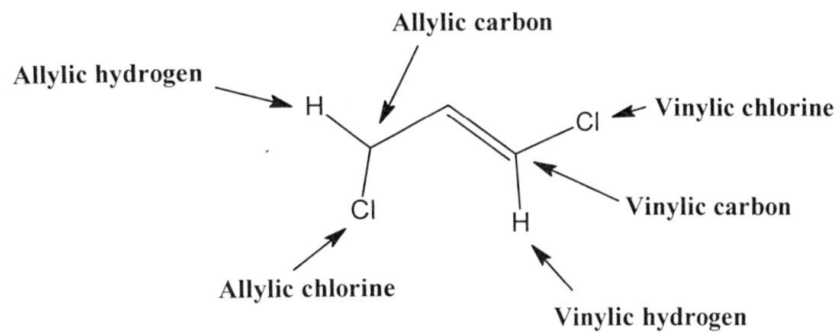

The term *"benzylic"* is similar to "allylic", but applies to an aromatic ring rather than to a C=C bond.

1. The **benzylic carbon is a carbon bonded with an sp^2-carbon of an aromatic ring.**

2. The **benzylic group** or a non-carbon atom is a group or atom **bonded with a benzylic carbon**.

Benzylic carbocations, free radicals, and carboanions are more stable than their allylic counterparts because the stabilizing conjugation involves more atoms.

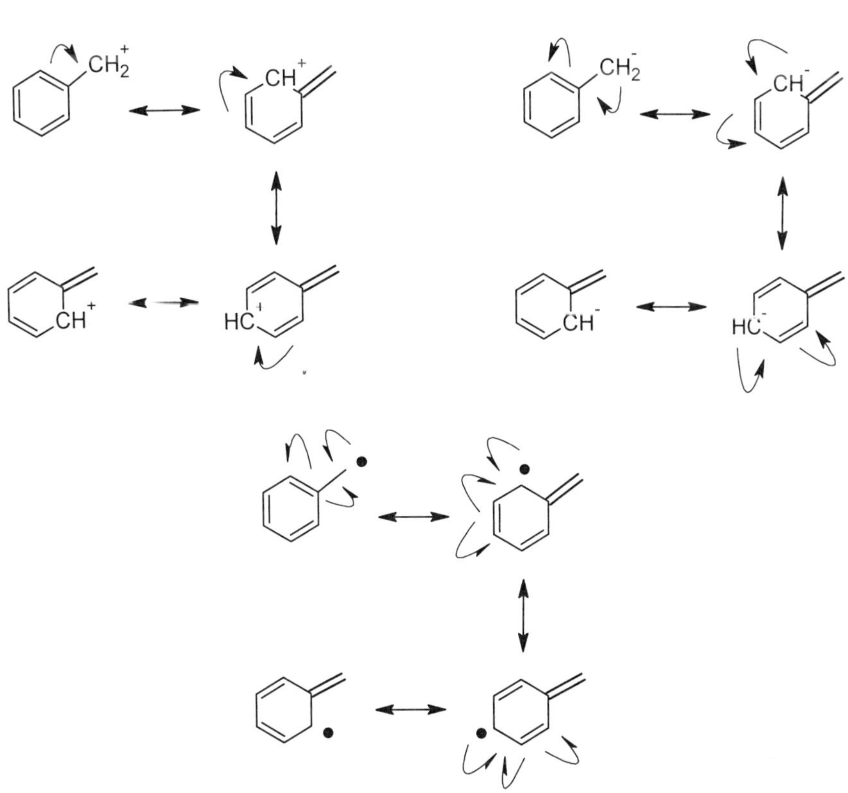

Examples of allylic and benzylic reactivity:

1. Primary allyl halides are about as active as tertiary alkyl halides in S_N1 reactions.

2. Free radical halogenation prefers to proceed to allylic or benzylic positions.

3. Benzylic carbons are more prone to oxidation than alkyl carbons.

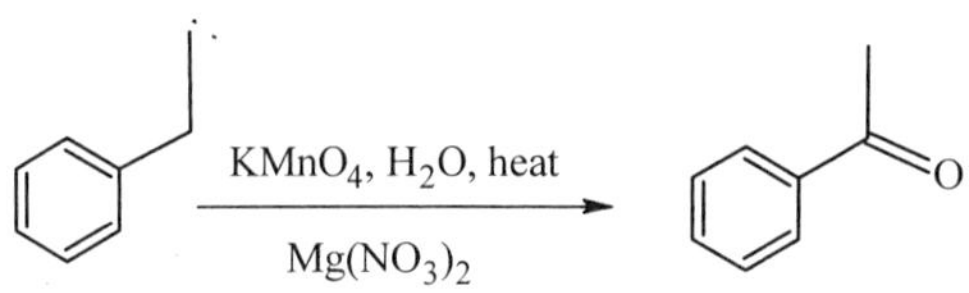

4. The industrial oxidative amination proceeds to the allylic position of propene and produced a very important monomer – acrylonitrile.

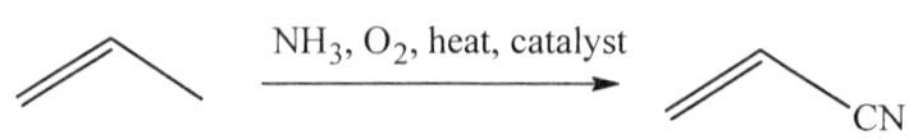

5. The conjugation not only stabilizes the reaction intermediate, but also can shift the reaction center. For instance, in an S_N1 reaction, the nucleophile attaches to a carbon, which is different from the carbon that lost the leaving group.

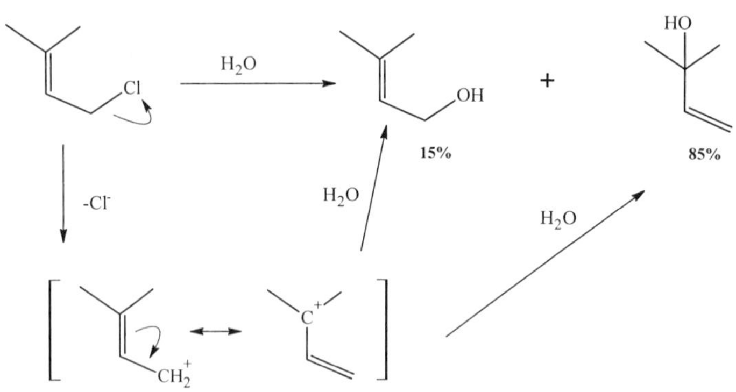

What can go wrong?

Relating of a structure of a 1,3-diene to the structure of the product of 1,4-addition is made very easy by the step-by-step procedure outlined in this chapter.

An 1,3-diene does not necessarily have the "1,3-" part in its chemical name.

Do not apply the *ortho-*, *meta-*, and *para-* terminology to the top of the chemical drawing, unless there is a substituent there, to which this terminology is related.

Do not apply the concept of *ortho-*, *para-*, and *meta-* directors and activators / deactivators to the incoming substituent in the reaction of aromatic electrophilic substitution.

The terms "allylic" and "vinylic" should not be confused. This chapter contains specific instructions of how to avoid it.

Pushing the limits

While higher reactivity of allyl and benzyl halides in the S_N1 and E1 reactions is now obvious for you, a closer look into the geometry of transition states will reveal that those halides are also more reactive in the S_N2 and E2 reactions. Moreover, abstraction of an allyl or benzyl *hydrogen* in the E1 and E2 reactions is preferred as well. The Diels-Alder reaction is a sneak peek into a huge variety of cycloadditions, including the Bergman cyclization, enabling the endiyne group of antibiotics and anticancer drugs.

Practice problems

1. Many years ago, a nation has found itself in the international isolation, and was cut off from imports of natural rubber, which is necessary for making cars and tractors for agriculture. A team of organic chemists led by Sergei Lebedev, has developed an alternative polymer with the structure shown below. This polymer, which has helped to stave off nationwide famine, was prepared by Na-promoted polymerization of a hydrocarbon. **Circle the positions of C=C bonds in the hydrocarbon - the starting material for polymerization.**

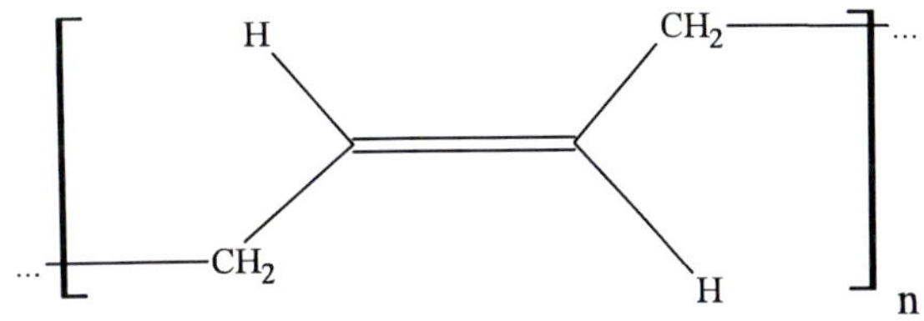

2. Circle all atoms of nitrogen that contain a lone electron pair, not participating in the cyclic conjugation.

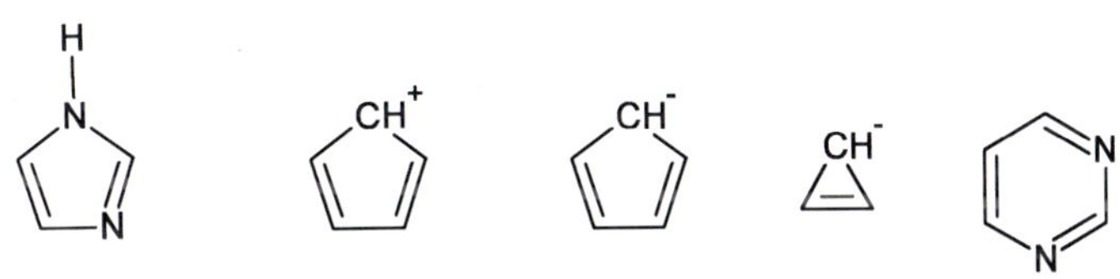

3. The reaction below shows formation of rhodopsin. **Which line on the spectral chart best represents UV-vis absorption of rhodopsin?**

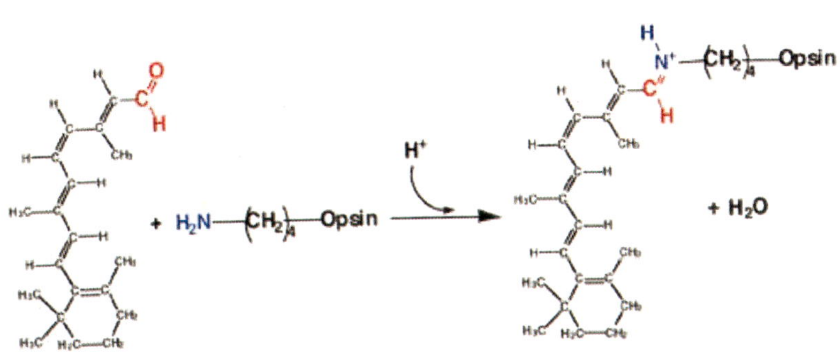

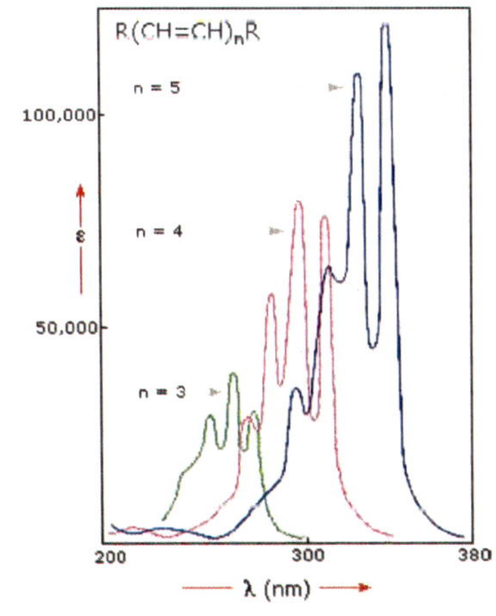

4. Circle all substituents that have a resonance electron withdrawing effect.

-Cl $-NH_3^+$ $-COCH_3$ $-NH_2$

-Br   -D $-CH_3$ -OH

5. Circle all hydrogens of methylbenzene that are available for substitution to produce the major reaction product(s) upon treatment with Br_2 / $AlBr_3$

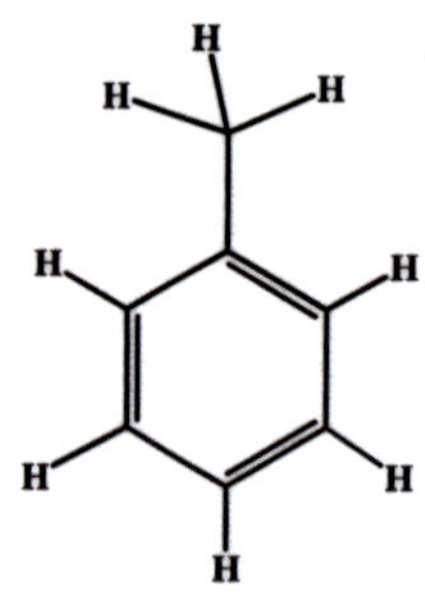

6. Circle all hydrogens of methylbenzene that are available for substitution to produce the major reaction product(s) upon treatment with Br_2 in the presence of light.

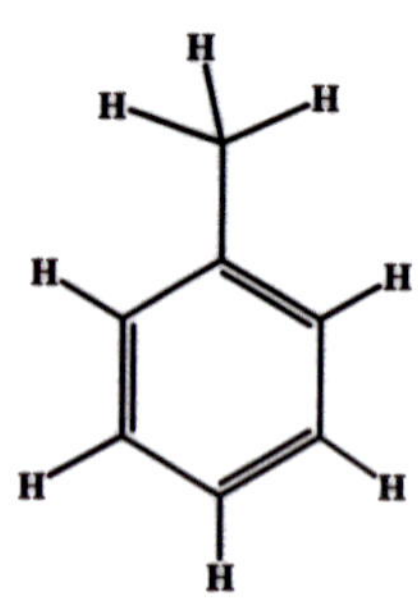

7. Circle the product of 1,4-addition of bromine to 1,3-butadiene.

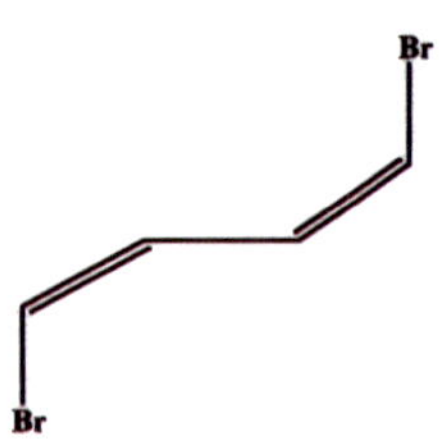
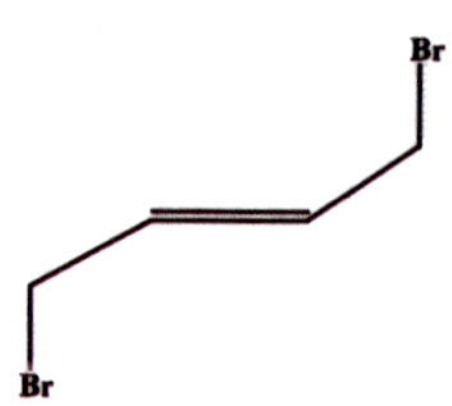
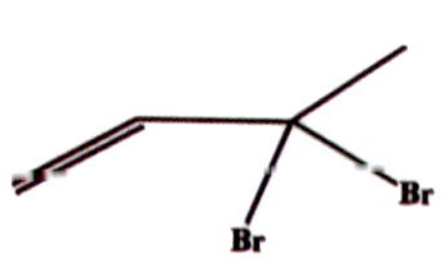
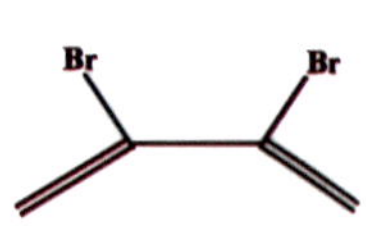
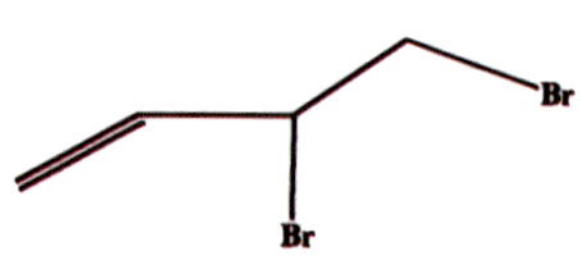

8. Circle the product of 1,2-addition of bromine to 1,3-butadiene.

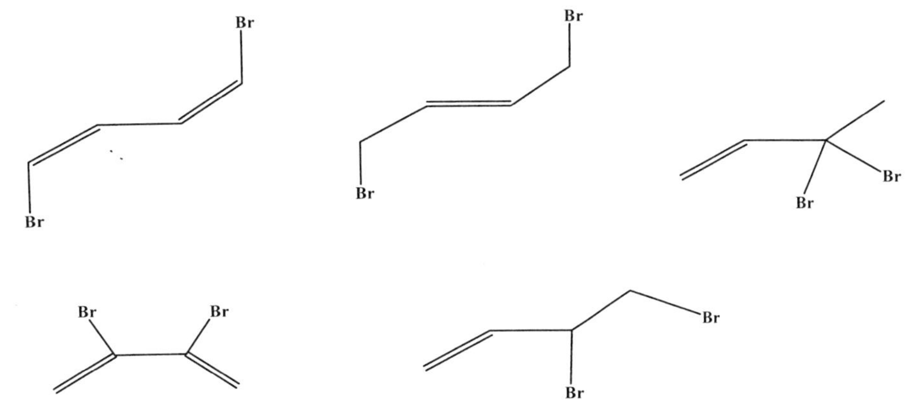

9. Electrophilic chlorination of acetylbenzene proceeds preferentially to the meta-position. **Why?**

A. The acetyl group is electron donor **B.** The acetyl group is electron acceptor

C. Cl is electron donor **D.** Cl is electron acceptor

E. Cl has three lone electron pairs

Making connections

Placing together double bonds lays groundwork for the concept of aromaticity. Aromatic electrophilic substitution is much easier to study by its comparison with electrophilic addition to alkenes. After this chapter, you are also ready to learn a variety of reactions of *aromatic nucleophilic substitution*. Understanding of the chemistry of vision is a well-deserved prize for understanding of molecular geometry, stereochemistry, and conjugation of C=C double bonds.

Chapter 7

Aryl halides,amines,and phenols.

The presence of halogens, amino- and hydroxy-groups in an aromatic ring (*aryl derivatives*) leads to two significant features .The aryl halide now has a leaving group for eliminations and nucleophilic substitutions.

1. Influence of the aromatic ring on the amino- and hydroxy-groups is so strong, that the hydroxy-derivatives of aromatic compounds are not considered as alcohols anymore. They are referred to as *phenols*.

For the E2 elimination in aryl halides, the most favorable conformation is not available. In addition, the reaction introduces to the product *(benzyne)* a carbon-carbon triple bond with a very strained geometry, which cannot be linear because it is a part of the ring.

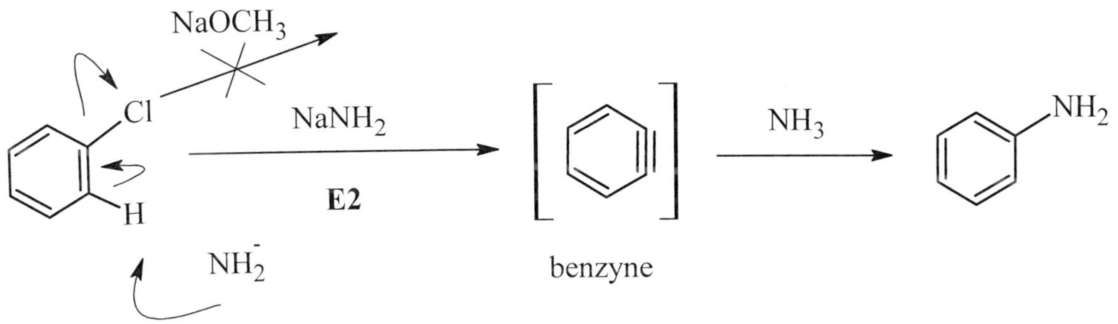

Nevertheless, the E2 reaction is possible, but it requires a very strong base (such as NH_2^-) or high temperature. The reaction product is so reactive, that it rapidly adds a nucleophile. The whole reaction sequence *(elimination followed by addition)* is a *nucleophilic aromatic substitution.*

The benzyne intermediate is so reactive, that the second addition step proceeds with virtually no regioselectivity. The telltale sign of this reaction mechanism is shift of the reaction center that is the leaving group leaves one carbon atom, and the nucleophile adds to another one.

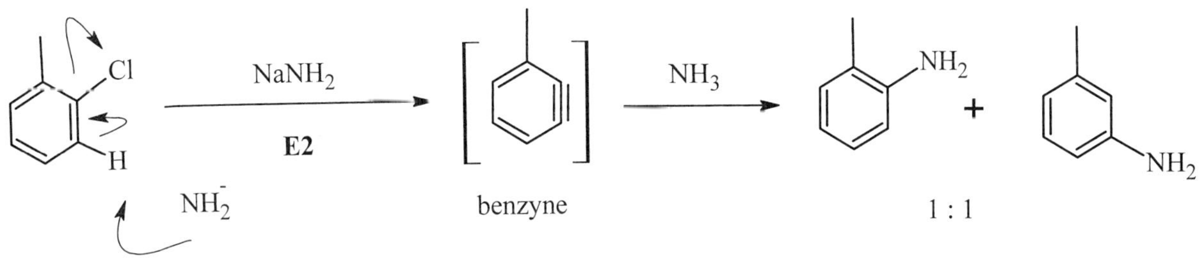

Contrary to the electrophilic aromatic substitution, the nucleophilic aromatic substitution can proceed through a variety of mechanisms. The next mechanism that we will discuss, is very similar to the mechanism of electrophilic aromatic substitution, but all charges and electron movements are reversed. Also, we need to note that in the electrophilic aromatic substitution, the leaving group is usually H^+, and in its nucleophilic counterpart, the leaving group is an anion, usually Hal^-.

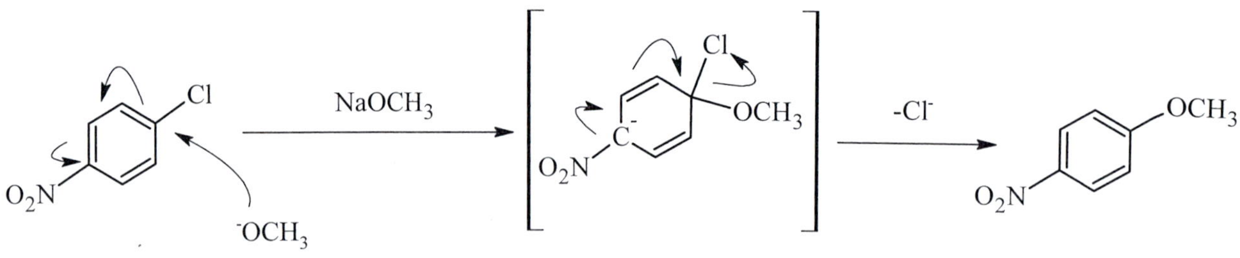

Meisenheimer's complex

However, formation of a Meisenheimer's complex is much less favorable than formation of a σ-complex in the reaction of aromatic electrophilic substitution. As a result, this type of aromatic nucleophilic substitution *(addition followed by elimination)* does not take place in the absence of activating groups. Following the complete reversal of the electrophilic substitution, one can figure out that strong resonance electron acceptors at the ortho- or para-position to the leaving group are activating groups, presence of which *is required for this type of substitution.*

Chemistry of phenols

Phenols contain a hydroxy-group at a carbon (sp^2) of an aromatic ring. Due to the strong interaction between the π-system of the aromatic ring and the lone electron pair of the -OH group, chemical properties of phenols significantly differ from the chemical properties of alcohols, where the hydroxy-group is bonded to an sp^3-carbon.

The strong resonance donating effect of the deprotonated -OH group stabilizes the conjugate base of a phenol (a phenolate), making phenols much stronger acids than alcohols (by about 7 orders of magnitude). The electrically neutral -OH group itself is also an excellent resonance donor, but not as efficient as the negatively charged phenoxide.

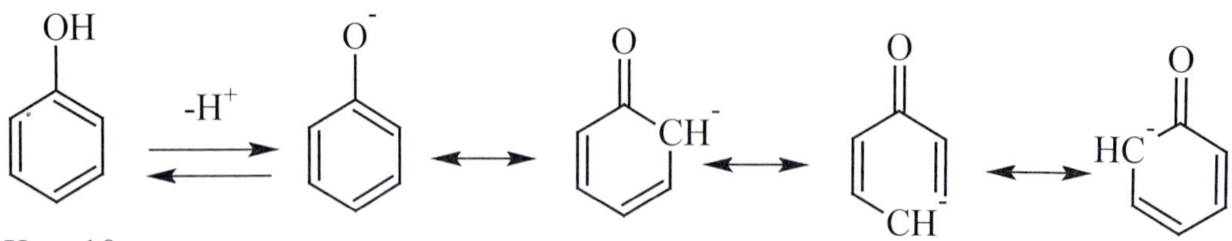

$pK_a = 10$

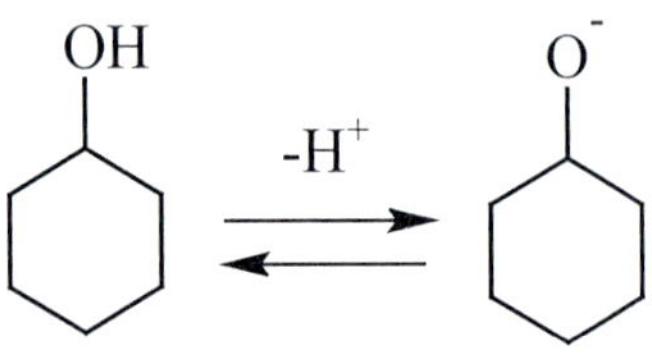

$pK_a = 17$

However, acidity of phenols is still about 5 orders of magnitude lower than that of carboxylic acids. Sodium hydrocarbonate, which is basic enough to deprotonate carboxylic acids, but not basic enough to deprotonate phenols, can be used to distinguish and, sometimes, separate those classes of compounds.

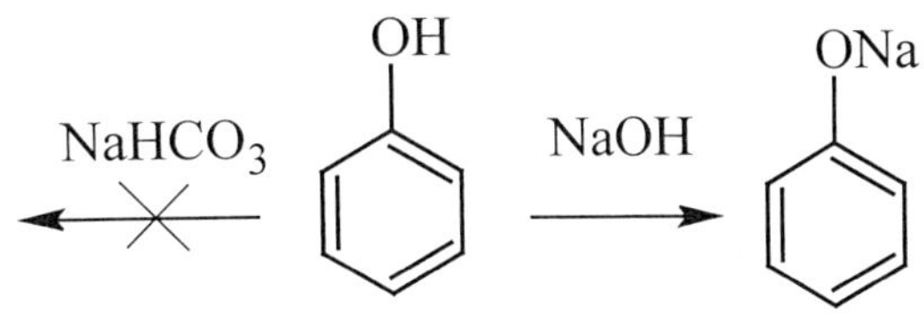

Compare:

$$RCOONa \xleftarrow{NaHCO_3} RCOOH \xrightarrow{NaOH} RCOONa$$

Conjugation with the aromatic ring stabilizes not only the negative charge, but also a free radical on the oxygen. As a result, the phenoxy-radicals are so much stable and less reactive than most other free radicals, they are unable to continue free radical chain reactions. Therefore, phenols are able to stop many oxidation reactions that proceed by the free radical chain mechanism.

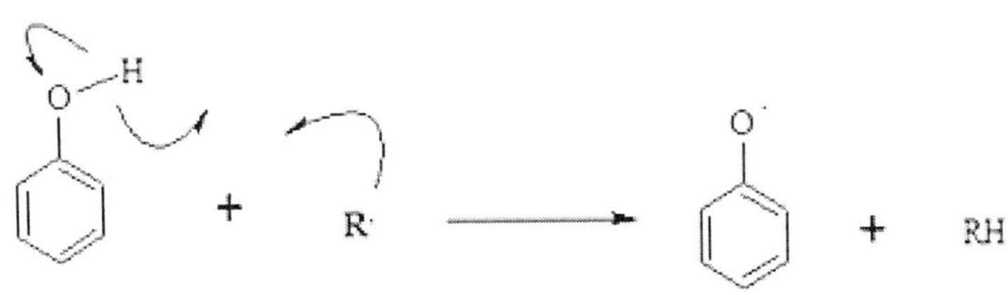

stable phenoxy-radical

In this role, phenols act as *antioxidants* with three main applications:

1. Nutritional antioxidants, suppressing harmful free radical side reactions of cellular respiration.
2. Food preservatives, suppressing free radical oxidation reactions of food spoilage.
3. Chemicals stabilizers, suppressing undesired free radical reactions of those chemicals, for example free radical polymerization of styrene.

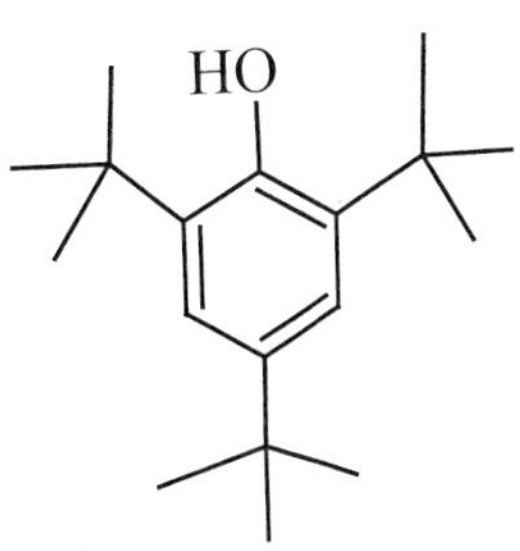

a food preservative

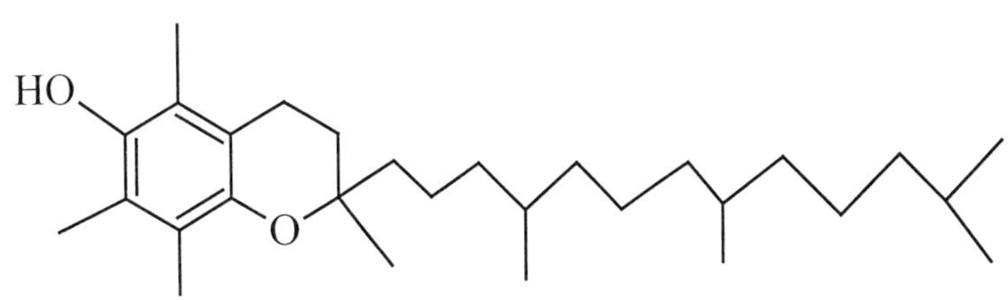

Vitamin E

While the aromatic ring significantly affects the attached –OH group, their interaction is reciprocal. As a strong resonance donor, the hydroxy-group activates the aromatic rings toward aromatic electrophilic substitution so much, that many of those reactions do not require a catalyst. For instance, phenolates can be carboxylated by a very weak electrophile – carbon dioxide. Interestingly, the reaction outcome depends on the cation of the base used to activate phenol.

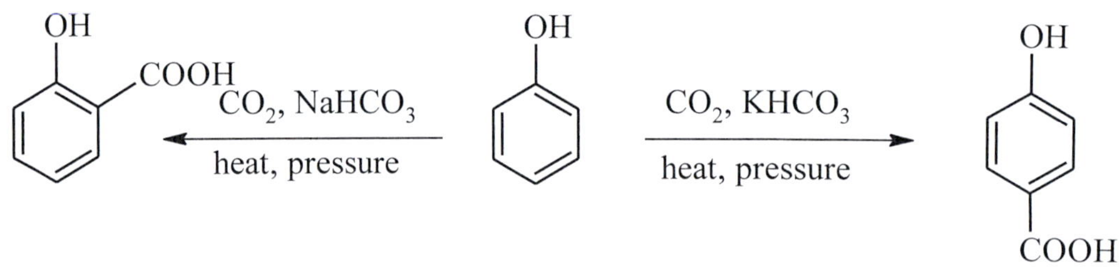

Salycilic acid — para-hydroxybenzoic acid

The smaller in size sodium cation coordinates with both carbon dioxide and the phenol hydroxyl better than a larger potassium cation. As a result, Na^+ guides the CO_2 molecule to the ortho-position of the hydroxy-group, leading to salicylic acid – the industrial precursor of aspirin. The higher affinity of Na^+ to lone electron pairs also accounts for the water-guiding function of Na^+ in the human body. In the presence of K^+, the reaction outcome shifts to para-hydroxybenzoic acid, whose esters are well known as preservatives (parabenes).

Phenols can be synthesized by the benzyne-type aromatic nucleophilic substitution from either chlorobenzene or benzenesulphonic acid.

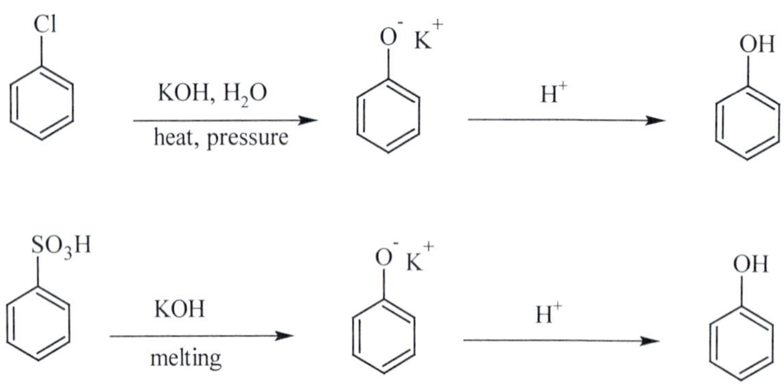

The modern industrial process for the production of phenols includes catalytic oxidation of isopropylbenzene (cumene), followed by an acid-catalyzed rearrangement, which is also the main industrial source of acetone. Phenols can be also prepared by the acidic cleavage of their ethers.

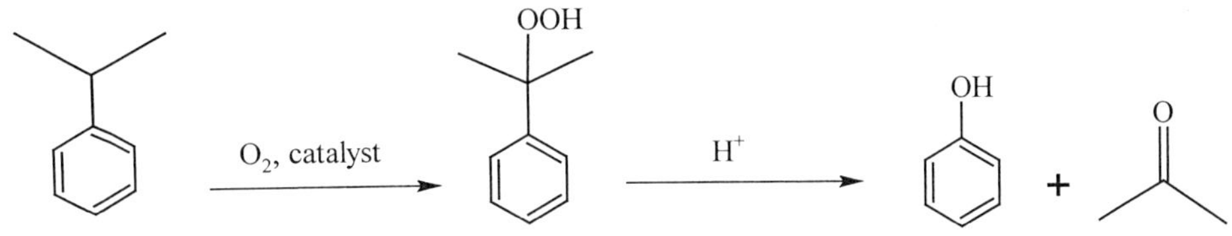

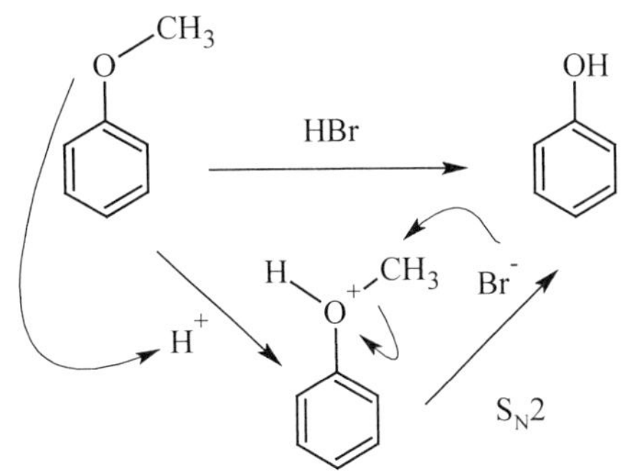

Chemistry of aromatic amines

The amino-group is even a stronger resonance electron donor than the hydroxy-group. However, it does not deprotonate as easily as the phenol hydroxy-group. This may explain why aromatic amines are still considered as amines, but phenols are not considered as alcohols.

Similarly to phenols, the interaction between the amino-group and the aromatic ring activates the ring toward electrophilic aromatic substitution and decreases basicity of the amine. As a consequence, the pK_a values of protonated aromatic amines fall well below 7. It means that in the human body aromatic amines are neutral, in contrasts with the positively charged aliphatic amines.

Perhaps the most significant difference in the chemical properties of aromatic and aliphatic amines is the ability of aromatic amines to form relatively stable *diazonium salts.*

Diazonium salts (or diazonium cations) are formed upon treatment of a primary amine with N_2O_3 usually generated *in situ* from a nitrite and an acid. However, due to high stability of the molecular nitrogen (N_2), it tends to break away, leaving behind a carbocation.

$$NaNO_2 \xrightarrow{H^+} HNO_2 \xrightarrow{-H_2O} N_2O_3$$

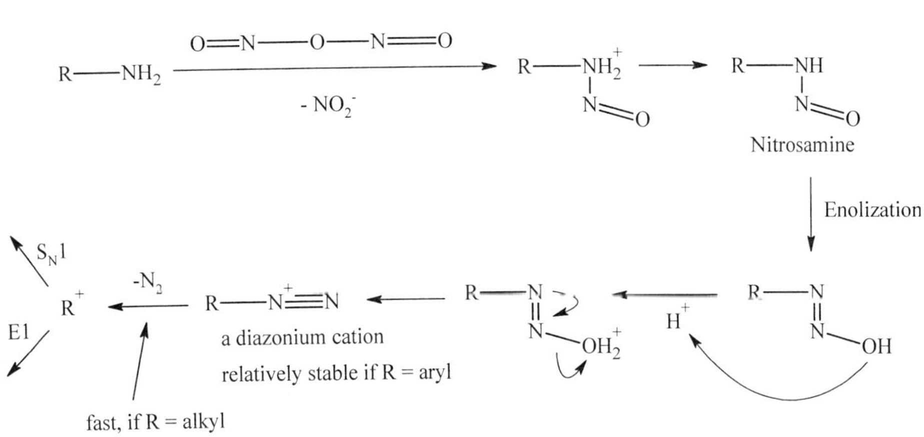

For aliphatic amines, the diazonium cation immediately decomposes, and the carbocation usually undergoes either S_N1 or E1 reaction. For aromatic amines, however, the diazonium cation is more stable in solution, and can be taken into further transformations. There are two reasons of the elevated stability of aromatic diazonium cations. First, they are stabilized by conjugation of the nitrogen-nitrogen triple bond with the aromatic ring. Second, evolution of N_2 leads to an unstable aryl carbocation.

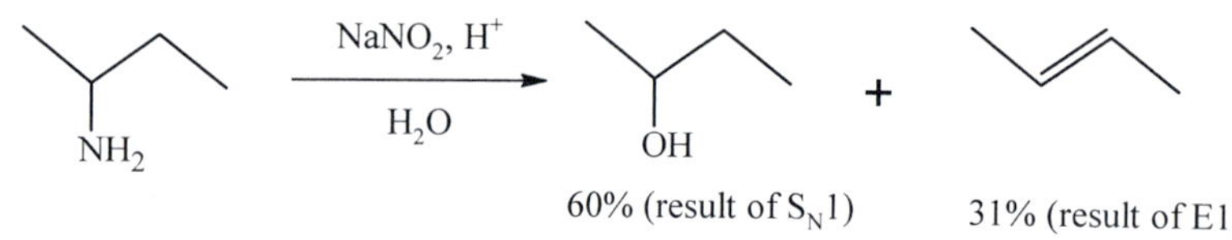

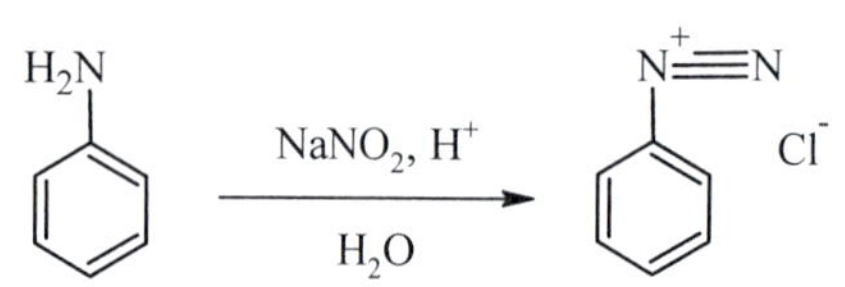

Once the aromatic diazonium cation is produced, it can undergo two types of reactions: *a) with loss of nitrogen; or b) without loss of nitrogen.*

a. The breaking away N_2 molecule makes an excellent leaving group, and this type of reactions occur in the presence of various nucleophiles. Multiple examples of this substitution are listed in any organic chemistry textbook.

b. The diazonium cation is a weak electrophile, able to react with activated aromatic compounds, such as phenols and amines. This reaction introduces a N=N bond between two aromatic rings, connecting them into an extended conjugated system (chromophore), and producing a wide variety of *azo-dyes*, for example, methyl orange.

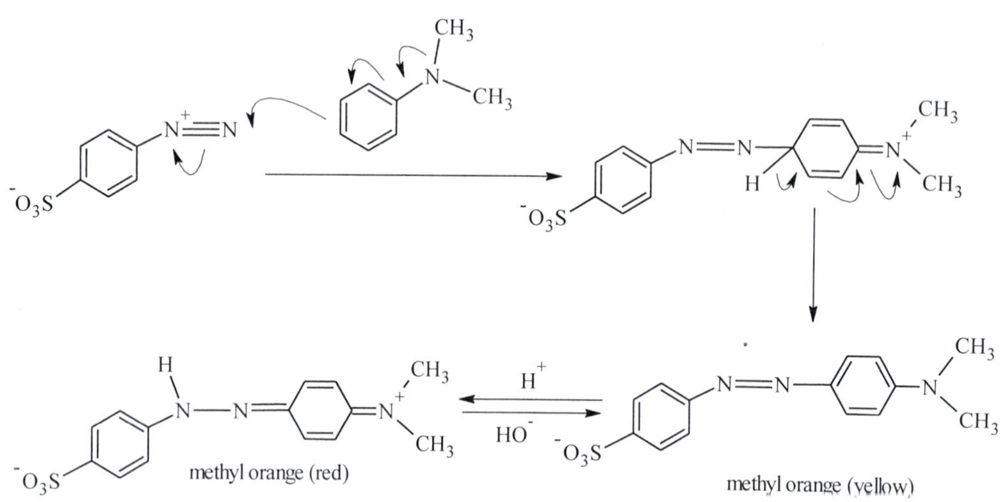

Protonation of the N=N bonds significantly affects the conjugated system and changes the pattern of conjugation, which, in turn, changes the dye's color. Thus, many azo-dyes are pH-sensitive and, therefore, widely used as pH-indicators.

What can go wrong?

A potential common confusion between nucleophilic and electrophilic substitution in the aromatic ring, especially in term of the leaving groups and directing substituents can be avoided by careful comparison of their mechanisms. The antioxidant activity of phenols by intercepting free radicals may seem contradictory to the previously learnt concept of reaction rates. We used to choose the reaction pathway through the most stable intermediate. However, if the intermediate is too stable, it may become too unreactive for the next step, which makes the whole reaction pathway unfavorable. Thus, switching a free radical reaction to the pathway through a stable phenoxy-radical suppresses the reaction.

Pushing the limits

Aromatic nucleophilic substitution is much more diverse than its electrophilic counterpart. In addition to the benzyne, Meisenheimer complex mechanisms and loss of nitrogen by diazocompounds, there are several more mechanisms of aromatic nucleophilic substitution. Also, reactions of diazocompounds without loss of nitrogen are not limited to the formation of azo-dyes.

Practice problems

1. The reaction of 1-chloro-4-methylbenzene with $NaNH_2$ / NH_3 leads to the formation of a mixture of 1-amino-4-methylbenzene and 1–amino-3-methylbenzene. **Presence of which reactive intermediate is consistent with such shift of the reaction center?**

   A. 4-Methylbenzyne B. A carbocation

   C. An anionic Meisenheimer complex

   D. A free radical E. A tetrahedral intermediate

2. Resveratrol (the structure shown below) found in red wine has been shown to be a potent antioxidant, significantly extending lifespan (by 10-50%) of many types of organisms by scavenging of harmful free radicals from mitochondria. Resveratrol quenches harmful free radicals by donating a hydrogen atom to them. **Circle all hydrogen atoms of resveratrol available for donation to the harmful free radicals.**

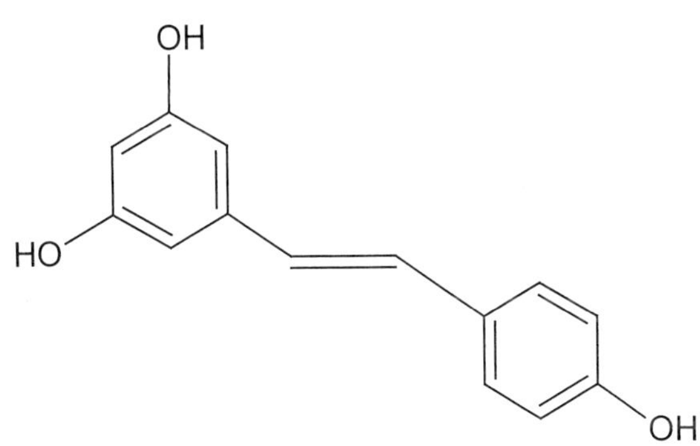

3. Introduction of strong electron acceptors to the molecule of chlorobenzene enables its reaction with ammonia at room temperature. **Presence of which reactive intermediate is consistent with this observation? Draw structure of this intermediate.**

   A. 4-Methylbenzyne B. A carbocation

   C. An anionic Meisenheimer complex

   D. A free radical E. A tetrahedral intermediate

4. 2,4,6-Trimethylphenol suppresses free-radical reactions by converting highly reactive free radicals to relatively stable free radicals unable to continue the free-radical chain oxidation. **Why are the**

phenoxy-radicals so stable? Choose the most important factor.

A. Resonance conjugation with the para-methyl group

B. Resonance conjugation with the ortho-methyl groups

C. Resonance conjugation with the π-bonds of the benzene ring, which delocalizes the free radical equally between five carbons of the ring and oxygen

D. Resonance conjugation with the π-bonds of the benzene ring, which delocalizes the free radical mostly between three carbons of the ring and oxygen

E. The reaction switches to a pathway through a stable carbocation

5. Which of the following compounds is a product of the reaction between 1-chloro-2,4-dinitrobenzene and hydrazine (N_2H_4)?

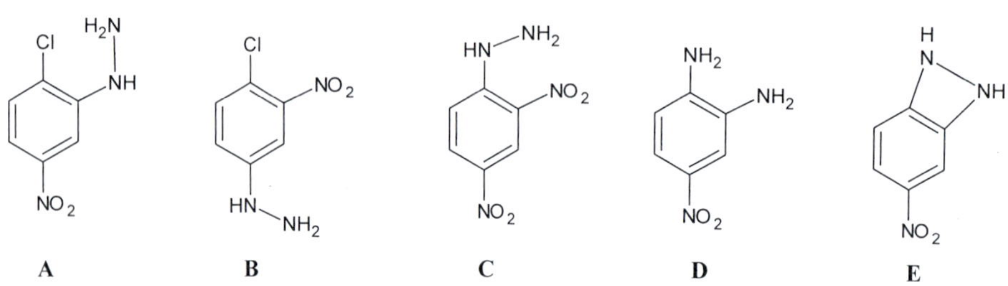

6. What is the source of two atoms of nitrogen in the diazonium salt formed from 4-nitro-aminobenzene and nitrous acid?

A. One nitrogen from the nitro-group, and another one from the amino-group

B. One nitrogen from the nitro-group, and another one from nitrous acid

C. One nitrogen from nitrous acid, and another one from the amino-group

D. Both nitrogens from nitrous acid

E. Both nitrogens from nitrogen-gas

7. Which of the following compounds will convert the diazonium salt to the dye "Methyl red" shown on the scheme?

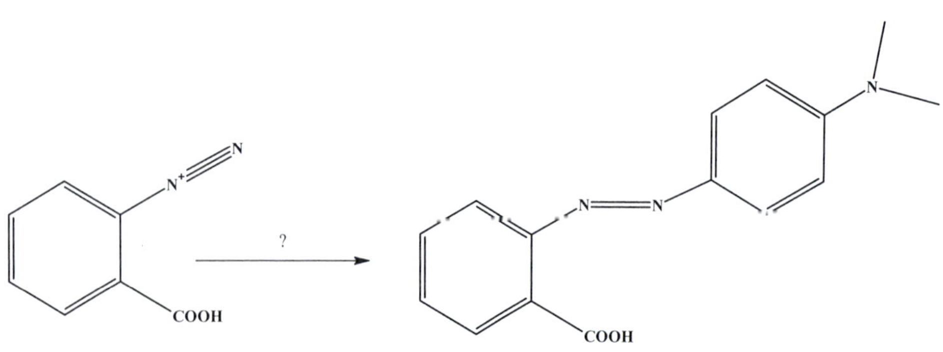

A. Aminobenzene

B. 4-(Dimethylamino)aminobenzene

C. Dimethylphenyl amine

D. 4-aminophenol

E. H^+ / HNO_2

8. **The reaction from question 7 belongs to which type?**

A. Aromatic electrophilic substitution

B. Aromatic nucleophilic substitution

C. S_N2 D. S_N1 E. Benzyne mechanism

9. All of the following compounds can be produced from diazonium salts. **In which case nitrogen gas does not evolve?**

A. Fluorobenzene B. Benzonitrile C. Benzene D. An azo-dye

E. Iodobenzene

10. An aqueous solution of an amine produced a white precipitate with aqueous Br_2. **Which of the following amines could it be?**

A. Methylamine B. Diethylamine C. 1-Aminohexane

D. 2-Aminohexane E. Aminobenzene

Making connections

The chemical activity of phenols as reducing agents and being stronger acids as alcohols allows for their abundance in nature (vitamins, biological signaling agents, components of wood) and variety of industrial applications. Combining your knowledge of phenols and amines with the material of the next chapter will introduce you to the chemistry of porphyrins, calixarenes, and phenol-formaldehyde plastics.

Chapter 8

Aldehydes and ketones

By this time, you are quite familiar with the chemistry of the C=C double bond. Now, we will see what happens if we replace one carbon with oxygen to produce the *C=O double bond,* which is also referred to as the *carbonyl group*.

The difference in electronegativities of carbon and oxygen make the carbonyl group very polar. First, it increases polarity of the whole molecule, making the carbonyl compounds more soluble in polar solvents such as water. Next, the electrostatic attraction of the partial electrical charges pulls the atoms together, making the carbonyl group shorter and stronger than the C=C double bond.

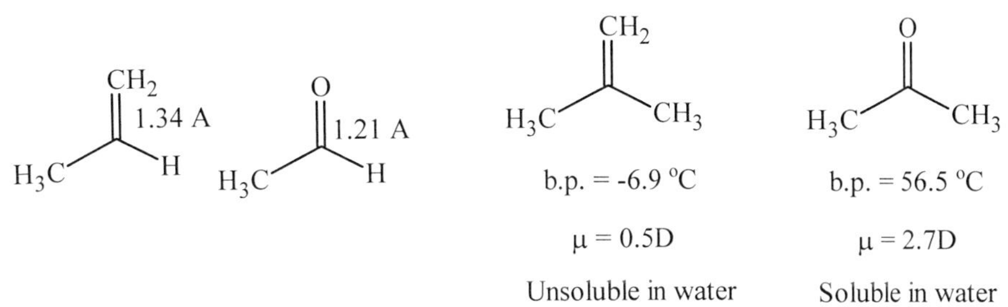

The partial negative charge developed on oxygen makes this atom basic and nucleophilic (able to react with acids and electrophiles), while the partial positive charge developed on carbon makes this atom acidic and electrophilic (able to react with bases and nucleophiles).

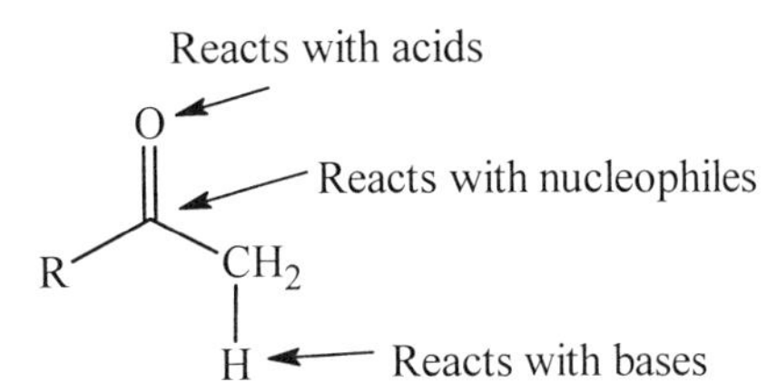

Aldehydes (R is H) or ketones (R is not H)

This chapter will focus on these two types of reactions of aldehydes and ketones. Note, that *neither in aldehydes nor ketones, the carbonyl group is attached to an atom with a lone electron pair*. This is because an atom with a lone electron pair tends to acquire another lone electron pair by using the two electrons of its covalent bond with the carbonyl group, breaking the bond and leaving the molecule. In addition, a lone electron pair significantly affects the neighboring carbonyl group by the resonance conjugation. As a result, the chemical properties of the carbonyl compound are changing so much, that it is considered as a different type of compounds, namely, a carboxylic acid or a carboxylic acid derivative, discussed in detail in the next chapter.

A lone electron pair next to the carbonyl group can be generated by abstracting a proton from the carbon next to the carbonyl group of an aldehyde or ketone. We will stir clear of it for now, and will completely devote Chapter 10 to this very important type of reactions.

Just like the C=C bond, the carbonyl group can undergo electrophilic addition, which is usually protonation. Electrophilic halogens are unlikely to attach to oxygen due to the weakness of the halogen-oxygen bond.

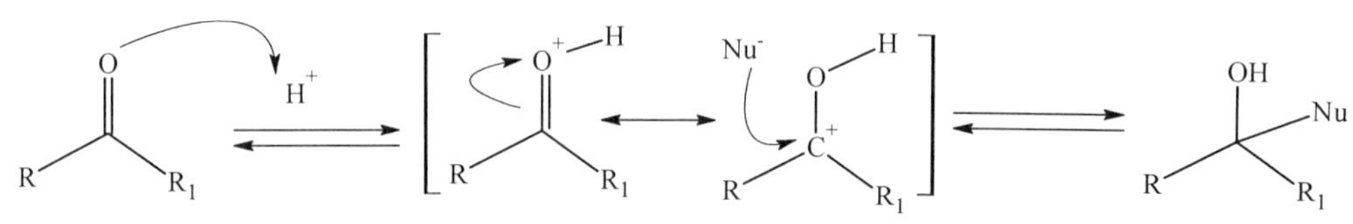

In contrary to the C=C bond, the carbonyl group can readily undergo nucleophilic addition due to the electrophilicity of the carbonyl carbon. Protonation of the intermediate alkoxide forms the same compound as produced by the electrophilic addition.

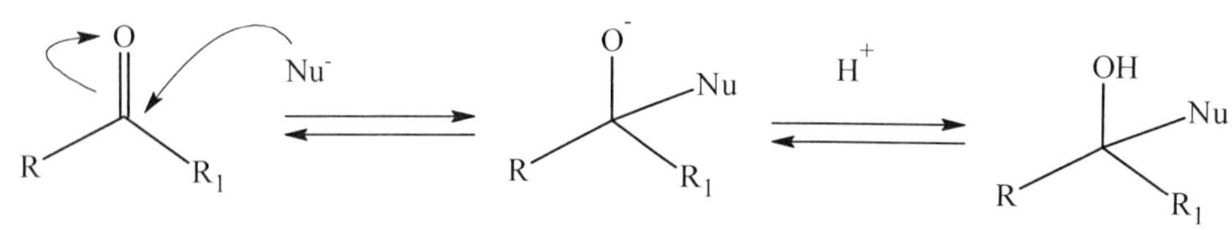

In this sense, the second step of the electrophilic addition to the carbonyl group can also be considered as nucleophilic addition to the carbonyl group activated by protonation.

The cyanohydrin synthesis (http://ochem.orgfree.com/cyan.html) is the simplest example of the nucleophilic addition to the carbonyl group.

Addition to the carbonyl group breaks the strong C=O bond, which makes the reaction thermodynamically unfavorable, that is the equilibrium is usually shifted back to the reactants. That is why aldehydes and ketones usually do not add water or HHal in a significant extent. The cyanohydrin synthesis is also reversible.

Addition to the carbonyl group proceeds in a significant extent, if the reaction is either irreversible, or followed by next steps, taking the system away from the unfavorable equilibrium.

Due to its irreversible nature, addition of a Grignard reagent to the carbonyl group is a very efficient method of building a new carbon-carbon bond.

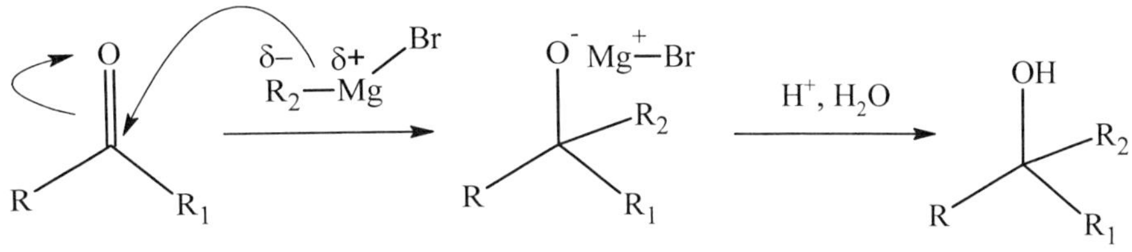

As far as the next steps following addition to the carbonyl group, it is often either elimination, or S_N1 substitution.

The first scenario is characteristic for the reaction of the carbonyl group with compounds of the general formula R-NH_2. Elimination of water from the product of addition produces a C=N bond, which is not as polar and strong as the C=O bond, but strong enough to shift the equilibrium toward the product.

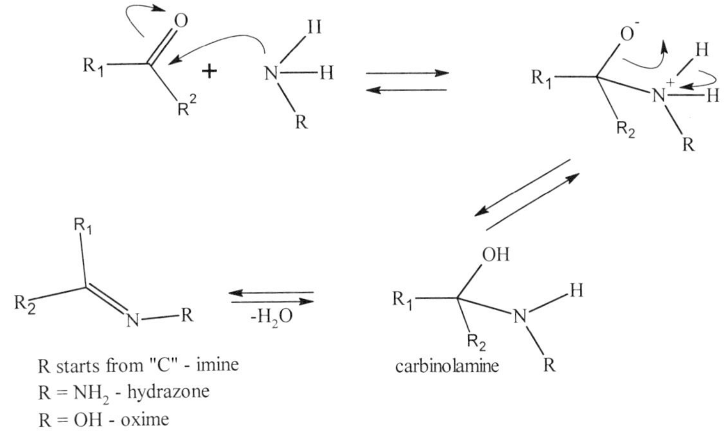

Examples of the reaction between aldehydes or ketones with the R-NH_2 compounds:

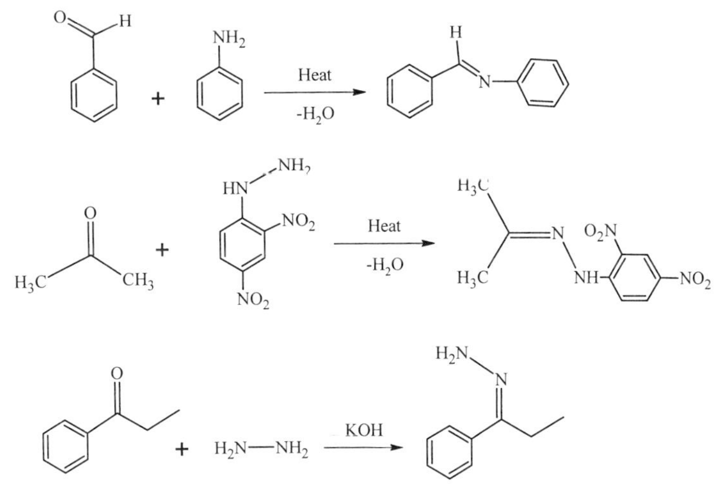

The second scenario is addition of alcohols to the carbonyl group (formation of hemiacetals), followed by the S_N1 reaction with another alcohol group (formation of acetals).

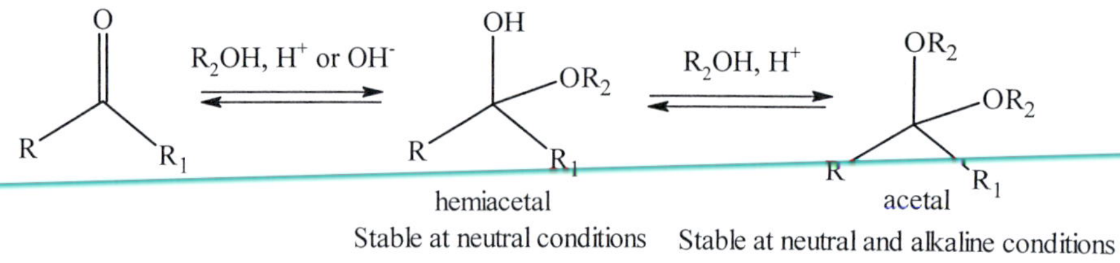

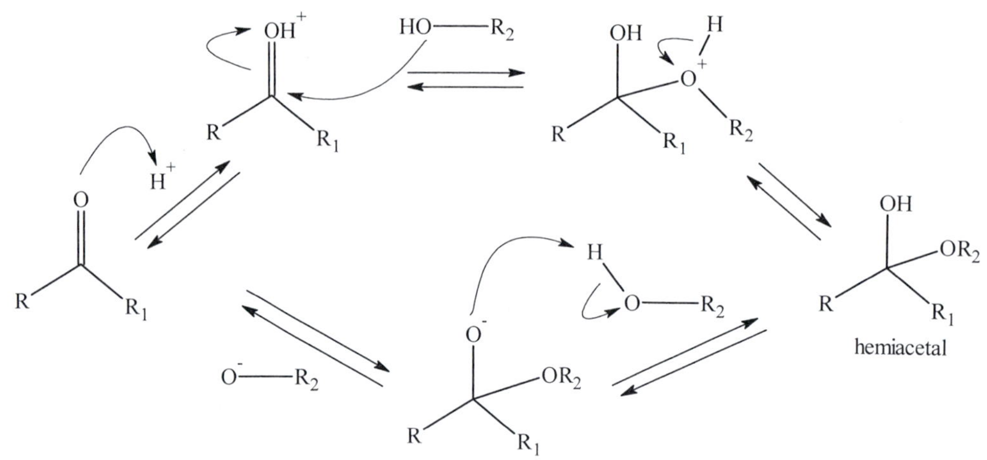

If the carbonyl group and the alcohol group belong to the same molecule, the reaction produces a cyclic hemiacetal. This is the chemistry behind the conversion of linear carbohydrates to their cyclic forms.

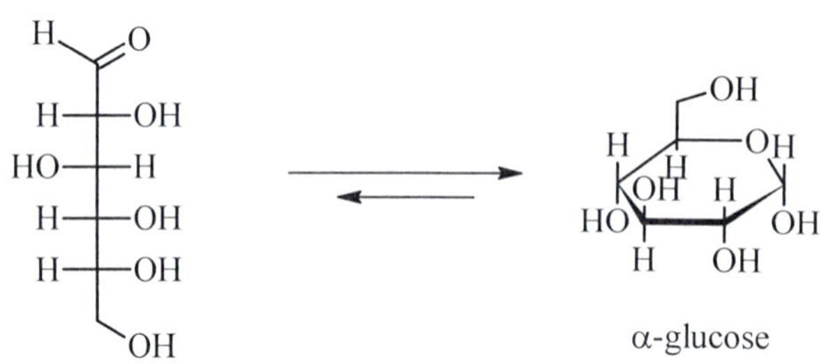

While hemiacetals can be formed or hydrolyzed back under either acidic or basic conditions, the S_N1 mechanism leading to acetals requires acidic conditions.

Examples of the syntheses of acetals

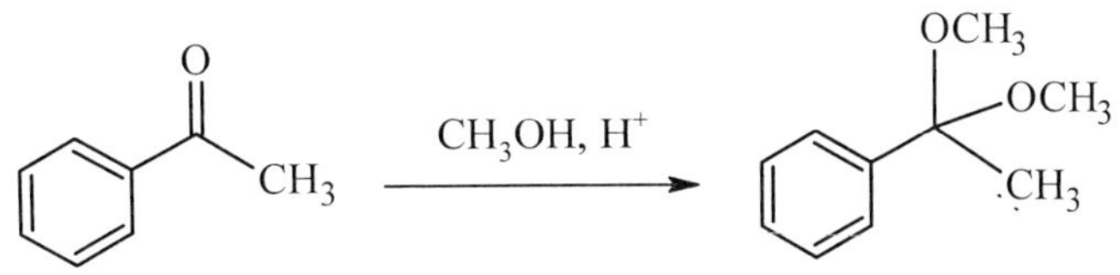

Methyl phenyl ketone dimethylacetal

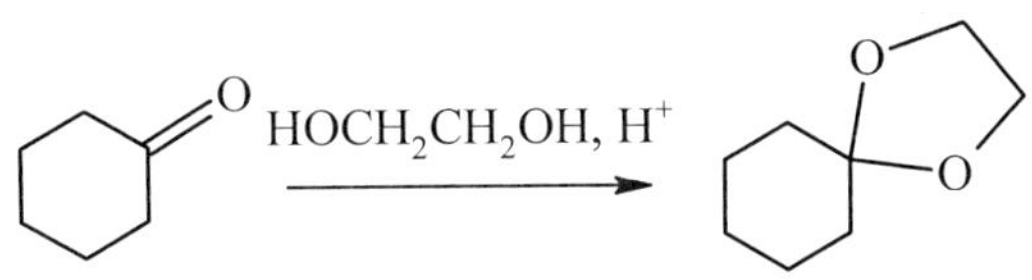

Cyclohexanone ethyleneacetal

In the second example, the alcohol group forming the hemiacetal and the alcohol group forming the acetal belong to the same molecule. This produces a cyclic acetal.

Hydrolysis of acetals also proceeds by the S_N1 mechanism and requires acidic conditions.

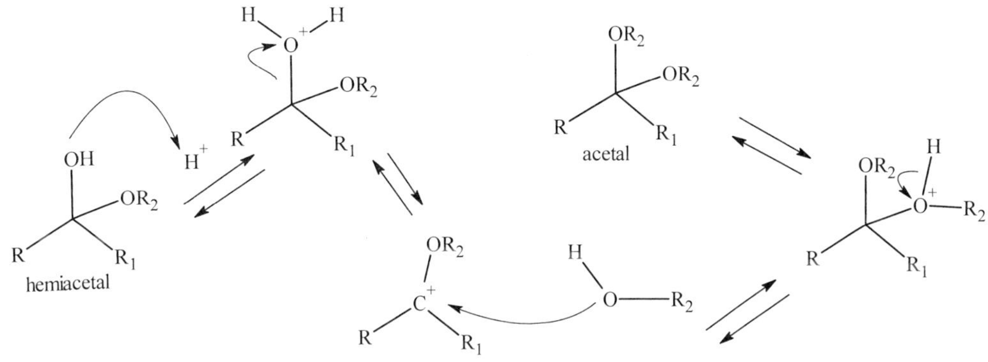

Therefore, *contrary to hemiacetals, acetals are stable under basic and neutral* conditions, which is not surprising, because acetals is a type of ethers. Understanding this will save you from memorizing in your biochemistry courses, which carbohydrates are reducing, and which are not. The reagent reduced by a carbohydrate is basic, and it oxidizes a carbonyl-containing fragment of a molecule. Therefore, if the structure of a carbohydrate contains at least one hemiacetal group, basic hydrolysis of this group will produce a carbonyl group, reacting with the oxidizing reagent. *A carbohydrate containing at least one hemiacetal group is a reducing carbohydrate (sugar). Non-reducing carbohydrates do not contain any hemiacetal groups.*

Due to its stability under the basic conditions and lability under acidic conditions, acetals are a very common protecting group that protects the compound from strong bases, and can be later removed under acidic conditions.

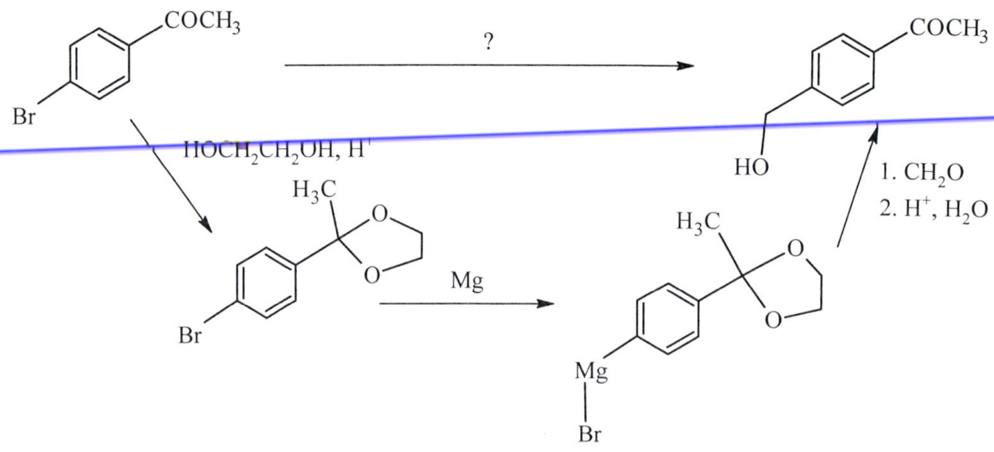

The acetal group of a natural compound, amygdalin, contained in certain types of almond, can undergo enzymatic hydrolysis, producing a cyanohydrin. Next, elimination (reversed nucleophilic addition to the carbonyl group) of the poisonous HCN from the cyanohydrin is responsible for toxicity of the types of almond, containing both amygdalin and the enzyme.

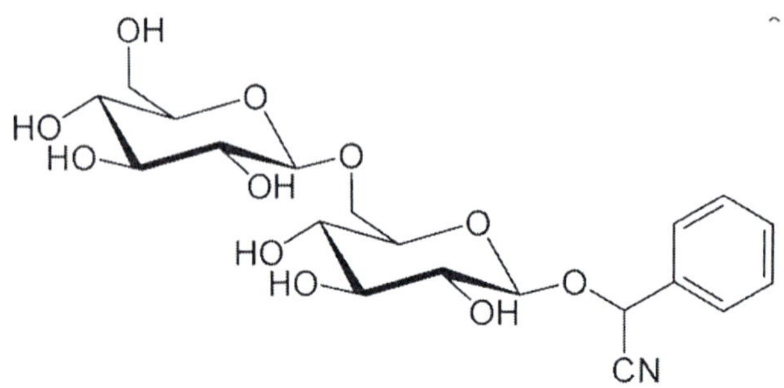

Amygdaline

Sometimes, the oxygen of the carbonyl group reacts with an electrophile, and the carbon of the carbonyl group reacts with a nucleophile synchronously. This happens in the reduction of aldehydes and ketones to alcohols by sodium borohydride.

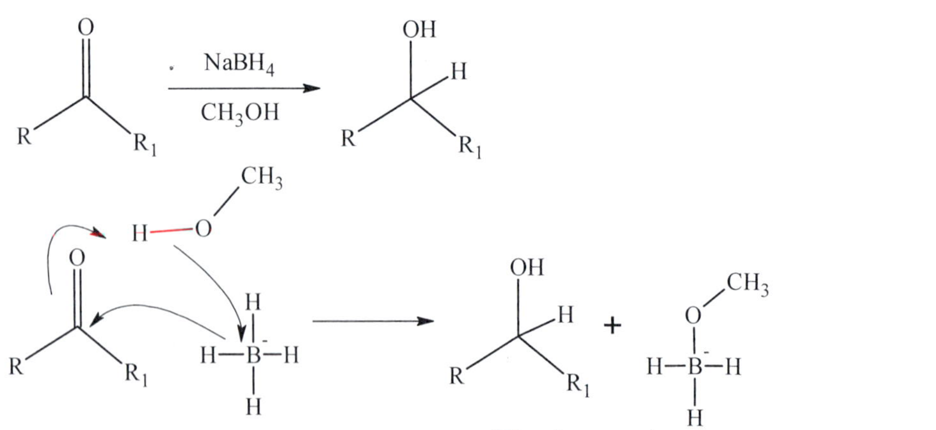

The protonated carbonyl group of aldehydes and ketones is electrophilic enough to act as an electrophile in the aromatic electrophilic substitution with activated aromatic compounds, such as phenols.

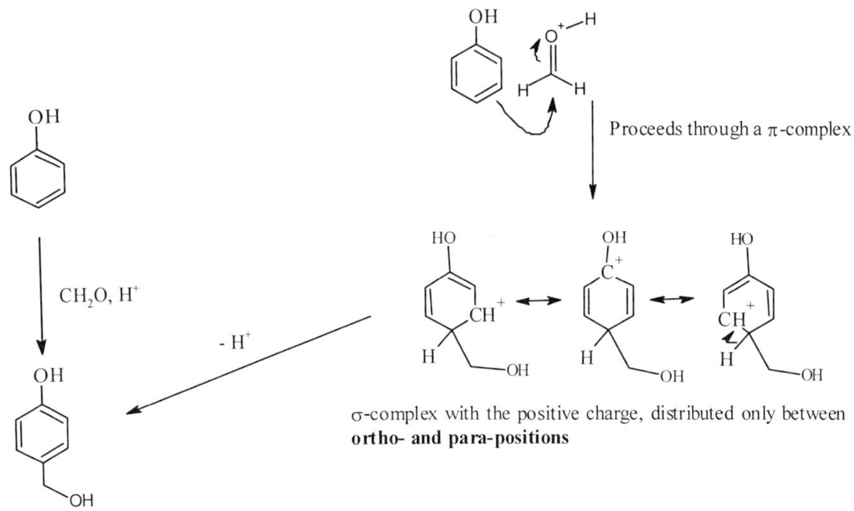

This reaction produces a benzylic alcohol and is called hydroxyalkylation.

Due to their ability to generate carbocations under acidic conditions, benzylic alcohols are good alkylation agents in the aromatic electrophilic substitution. Therefore, hydroxyalkylation is often accompanied with alkylation, leading to products of multistep reactions. For instance, the infamous *Bisphenol A* used in the production of plastics, can be prepared by condensation of phenol with acetone.

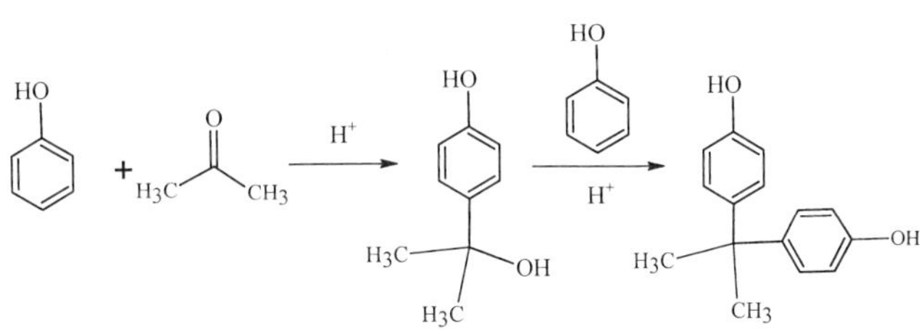

Bisphenol A

Similarly, condensation of phenol with methanal (formaldehyde) cross-links the molecules of phenols at the *ortho-* and *para-*positions by CH_2-groups, leading to a variety of *phenol-formaldehyde resins.*

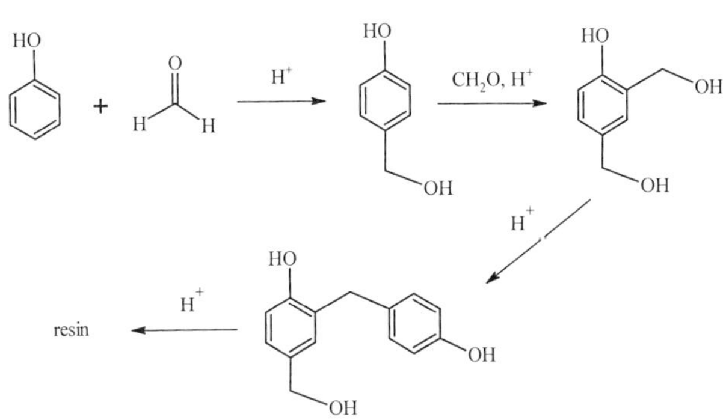

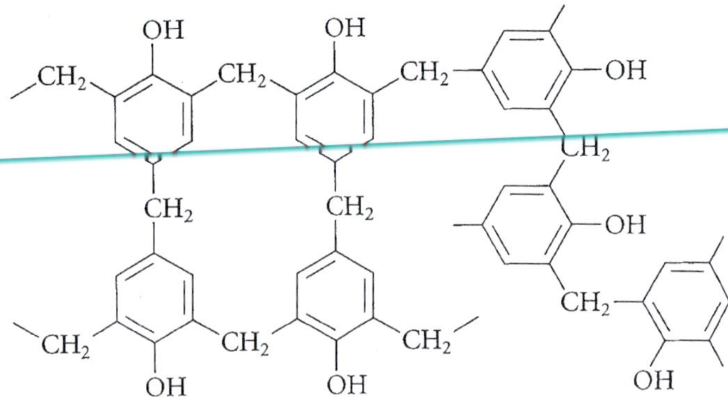

Structural fragment of a phenol-formaldehyde resin

Things becomes even more interesting, if the *para*-position of phenol is blocked by a bulky tert.butyl group, so the cross-linking at this position is disabled. Cross-linking at the ortho-positions only causes the molecule to bend and, finally, form a cycle, which is most often comprised by four or eight benzene rings.

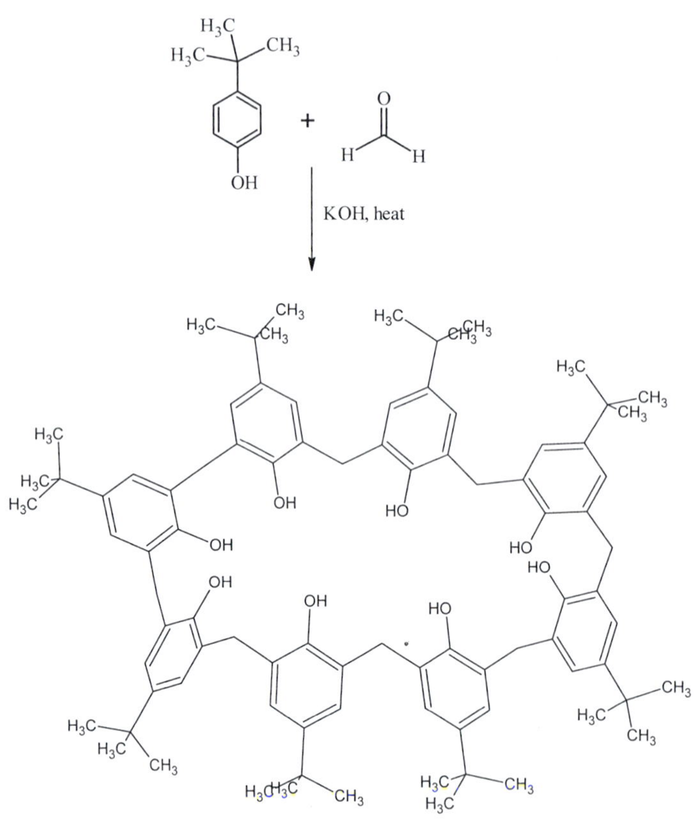

Synthesis of a calix[8]arene

Due to the presence of unique structural features (both polar and non-polar fragments, conformational flexibility, internal cavity, internal system of hydrogen bonds, about 1 nm size), calixarenes are organic nanoparticles able to form complexes with metals and small molecules. Calixarenes are popular building block in nanotechnology for constructing selective sensors, gas-separating membranes, heavy metal scavengers, catalysts, drug-delivery vehicles, etc.

Similarly, condensation of aldehydes with pyrrole (an aromatic molecule about as active toward electrophilic substitution as phenol), produces another molecule with a large cycle (macrocycle), containing four pyrrole rings. Oxidation of that product by air results in a flat aromatic system with a continuous conjugation of π-orbitals (porphyrin).

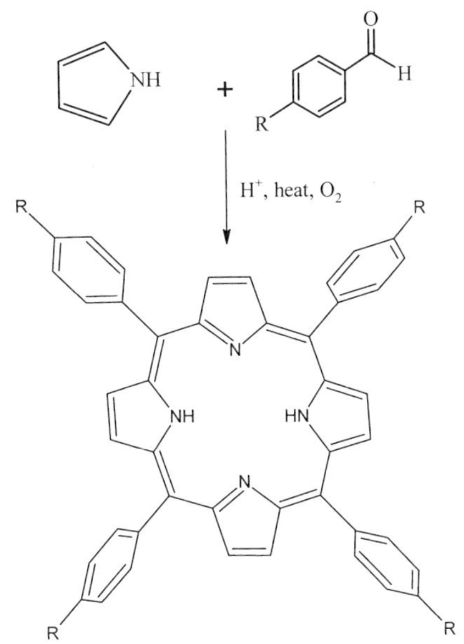

Synthesis of tetraarylporphyrins

The porphyrin core readily forms salts with Mg, making up a metal-binding site of chlorophyll, with Fe, making up a metal-binding site of hemoglobin, and with other metals. Synthetic porphyrins and their metal salts are incredibly useful as catalysts, agents for the photodynamic therapy of cancers and infections, components of solar cells, lasers, and other advanced devices.

In aldehydes and products of addition to the carbonyl group, abstraction of the aldehyde hydrogen in the form of H^- generates a positive charge at the adjacent carbon, which is stabilized by the conjugation with the oxygen's lone electron pair. Therefore, aldehydes are more prone to oxidation than ketones. The aldehyde group is readily oxidized to the carboxylic group by a variety of oxidizing agents. One of such reagents is Tollens' reagent ($[Ag(NH_3)_2]OH$), producing a silver mirror when reacts with an aldehyde group of reducing carbohydrates.

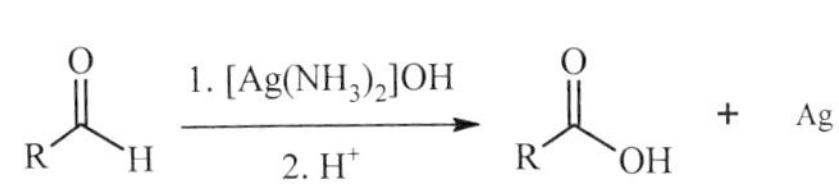

What can go wrong?

Students are often having trouble with the cyclic acetals and hemiacetals. There is no new chemistry there. The only difference between cyclic acetals and non-cyclic acetals is that in cyclic acetals, *both alcohol groups involved on the different reaction steps belong to the same molecule.* The only difference between cyclic hemiacetals and non-cyclic hemiacetals is that in cyclic hemiacetals, *both carbonyl and alcohol groups involved in the synthesis, belong to the same molecule.*

The hydrogen of the aldehyde group is usually so noticeable in the written structure, that it is tempting to consider it acidic. The aldehyde hydrogen is much less acidic than the hydrogen at the carbon next to the carbonyl group. You prior knowledge of acids and bases is sufficient to figure out why.

Pushing the limits

The carbonyl group is polarized with a partial positive charge on carbon, and partial negative charge on oxygen. However, when the carbonyl group absorbs UV-light, the electrons are re-organized so that in the excited state, carbon ends up to be more negative, and oxygen – more positive. This drastically changes chemistry of aldehydes and ketones, enabling the synthetic pathways unavailable in the ground (not-excited) state.

Practice problems

1. **The following reaction sequence converts an alcohol to a cyanohydrin by oxidation to an aldehyde followed by addition of HCN. Circle the hydrogen in the final product that was a part of the aldehyde group at the intermediate step.**

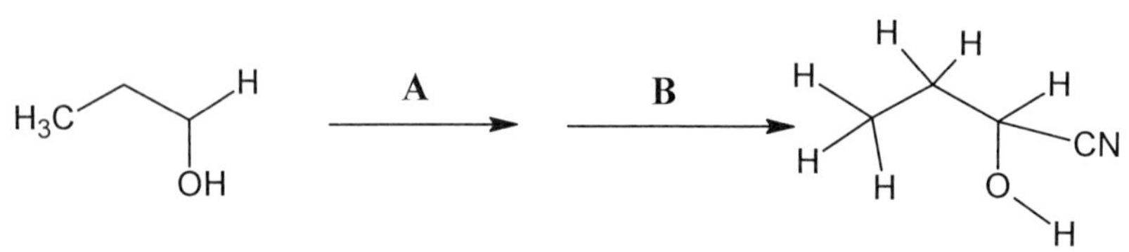

2. **Circle all parts of the acetal molecule that originated from an alcohol in the following synthesis:**

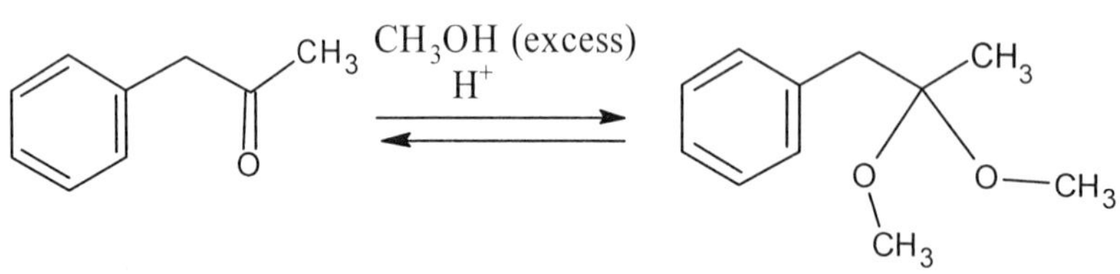

3. **Circle the least reactive compound in a highly alkaline (basic) aqueous solution.**

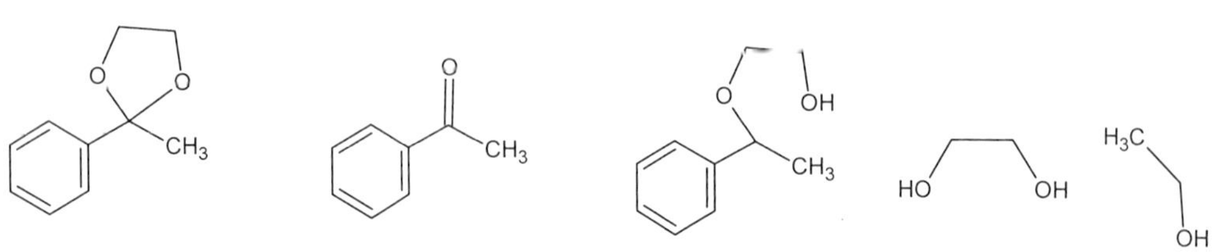

4. **Circle the reagent(s) used to remove the acetal protection from a ketone.**

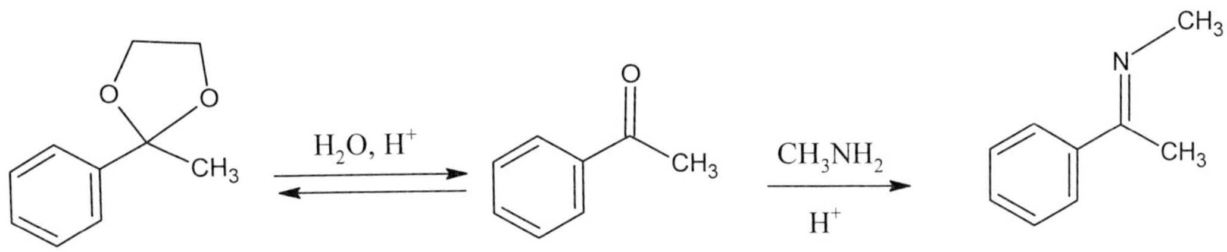

5. **Circle the methyl group that originated from methylamine**

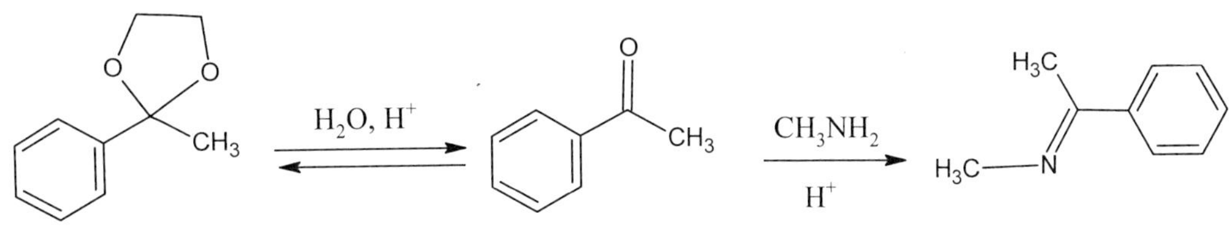

6. **Triphenylcarbinol can be synthesized from diphenylketone and phenyl magnesium bromide. What is the source of phenyl groups in triphenylcarbinol?**

A. One phenyl group from diphenylketone, two – from phenyl magnesium bromide

B. One phenyl group from phenyl magnesium bromide, two – from diphenylketone

C. One phenyl group from diphenylketone, two – from the solvent

D. Three phenyl groups from diphenylketone

E. Three phenyl groups from phenyl magnesium bromide

7. **Explain, why the following synthetic route will not work:**

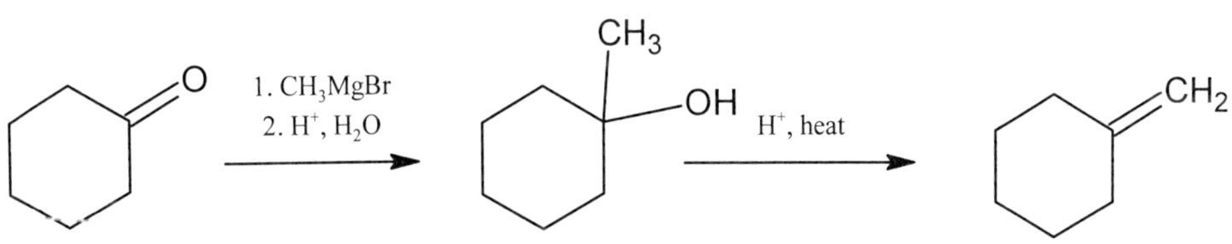

A. The Grignard reagent will not add to the ketone

B. The Grignard reagent will not add to the alcohol

C. The Markovnikov rule will lead to a different product

D. The Zaitsev rule will lead to a different product

E. The leaving group in the alcohol is too bad

8. **Which of the following compounds is an acetal?**

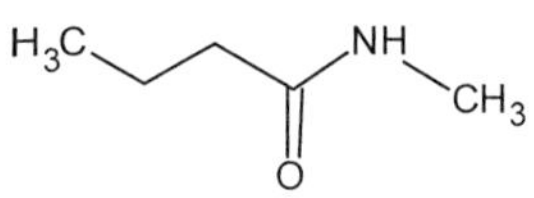

A

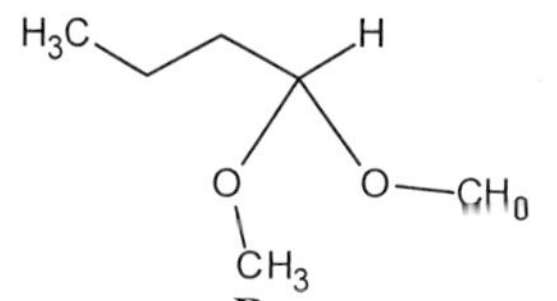

B

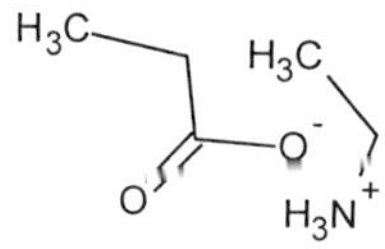

C

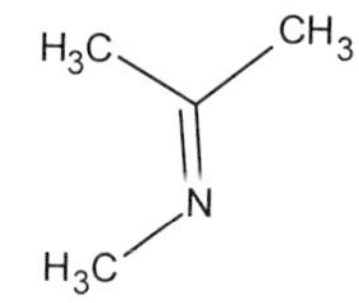

D

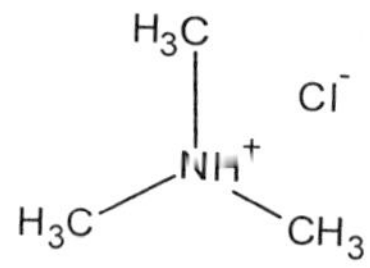

E

9. **Which reagents will produce the compound you selected for question 8?**

A. Butanoic acid and methylamine

B. Butanal and methanol

C. Propanoic acid and ethylamine

D. Propanone and ethylamine

E. Trimethylamine and HCl

10. **What is the role of acid in the mechanism of the acid-catalyzed synthesis of hemiacetals from aldehydes?**

A. Activation of the carbonyl group only

B. Activation of the nucleophile only

C. Activation of the carbonyl group and the leaving group

D. Activation of the carbonyl group and the nucleophile

E. Activation of the nucleophile and the leaving group

Making connections

It is so much easier to compare chemistries of the C=O group and the learnt C=C group than learning everything from scratch. Taking some time to review addition to the C=C group, eliminations and substitutions will save you much more time on learning aldehydes and ketones.

This material will prepare students for all remaining chapters of the course, especially, chemistry of enolates. Chemistry of acetals and semiacetals will tremendously help to understand cyclic forms of carbohydrates in the biochemistry courses.

Chapter 9

Carboxylic acids and their derivatives

In the previous chapter, we discussed chemistry of the carbonyl group bonded with either carbon or hydrogen (atoms with no lone electron pairs). Placing an atom with at least one lone electron pair next to the carbonyl group makes a profound influence on its chemical behavior. First, conjugation of the lone electron pair with the carbonyl group stabilizes the compound and *reduces its reactivity toward reactions where the C-L bond does not break (enolization, deprotonation, and addition to the carbonyl group)* as shown on the figure below.

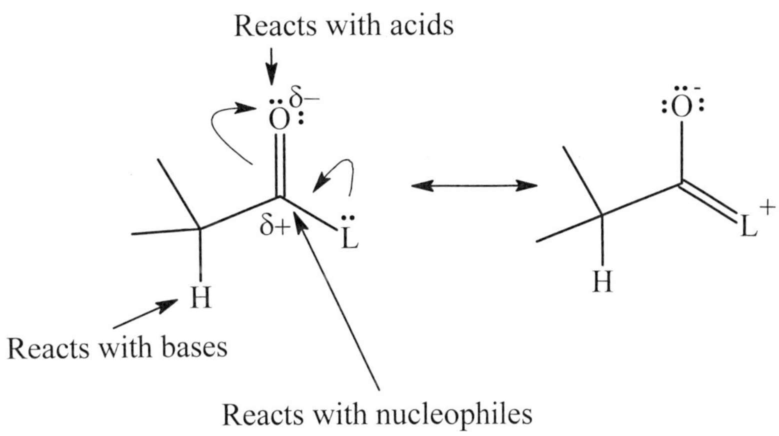

Second, presence of lone electron pairs is characteristic for more electronegative atoms that make much weaker bases and, therefore, much better leaving groups than carboanions. As a consequence, *such compounds are likely to undergo reactions of substitution (= addition + elimination)* as opposed to aldehydes and ketones (lacking good leaving groups), which undergo just *addition*.

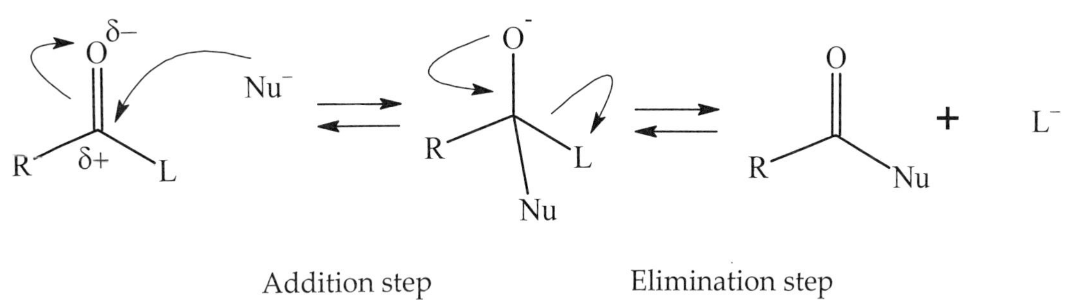

This type of substitution is termed *Acyl nucleophilic substitution*. The part of the molecule of a carboxylic acid derivative that is not the leaving group, is termed *Acyl*.

A special case of L is –O⁻, whose electron pair is evenly distributed between two electronegative oxygens, which makes the anion (termed *carboxylate*) much more stable than alkoxides and most phenolates, and makes the conjugate acid (termed *carboxylic acid)* stronger than alcohols and phenols (pK_a are usually in the range 3 – 6).

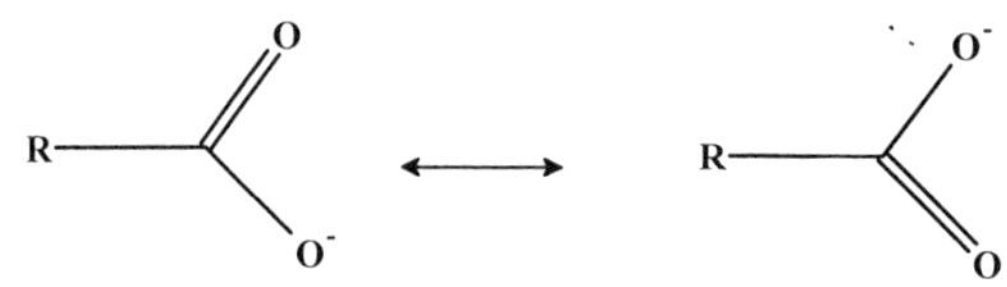

Carboxylic acids can be produced by substitution of X with –OH that may come from either water or its conjugate base (OH^-) in a reaction that belongs to the type of hydrolysis. Due to the common ability to produce a carboxylic acid or its salt, this type of compounds are considered as carboxylic acid derivatives. By definition, *carboxylic acid derivatives are compounds that produce a carboxylic acid or its salt upon hydrolysis.* Not all carboxylic acid derivatives consist of an acyl and a leaving group. Generally, a carboxylic acid derivative can be identified by the presence of a carbon connected by three bonds to an atom with at least one lone electron pair. Those three bonds can come in various sets (three single bonds, or one double bound and one single bond, or one triple bond), and can connect to one or more atoms with a lone electron pair. Below are some examples of carboxylic acid derivatives. Carboxylates are not always considered as carboxylic acid derivatives.

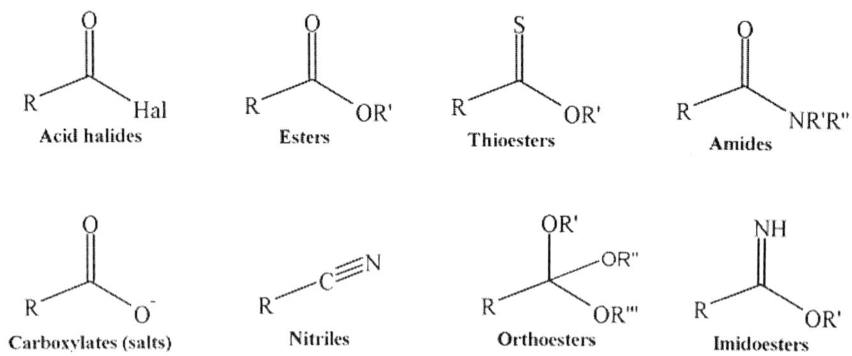

In this book, we will limit our consideration only to the carboxylic acid derivatives consisting of an acyl and a leaving group. While the relative reactivity of carboxylic acid derivatives depends on many factors, usually it can be easily predicted by comparing basicity of the leaving group (L):

The lower the basicity of L, the more reactive the carboxylic acid derivative is.

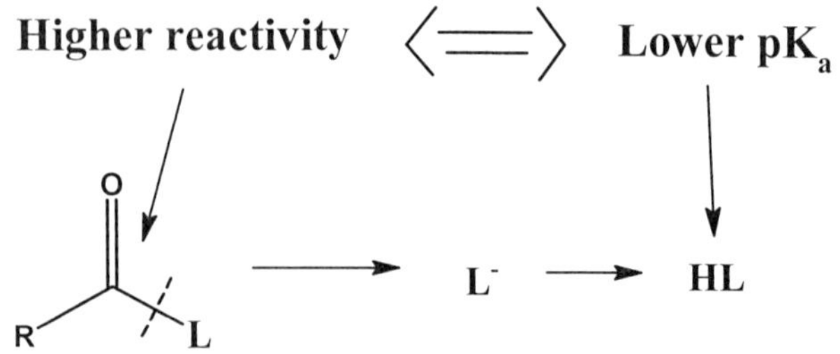

The relative reactivities of most common carboxylic acid derivatives have been summarized in the following table.

Carboxylic acid derivative	Leaving group L	pKa of HL
Acid chlorides *(Most reactive)*	Cl^-	-7
Anhydrides	Carboxylate	5
Esters	Alkoxide	15
Amides *(Least reactive)*	NH_2^-	36

Reactivity of a carboxylic acid derivative not listed in this table can be easily assessed just by looking up basicity of its leaving group.

Trans-esterification is the simplest example of acyl nucleophilic substitution.

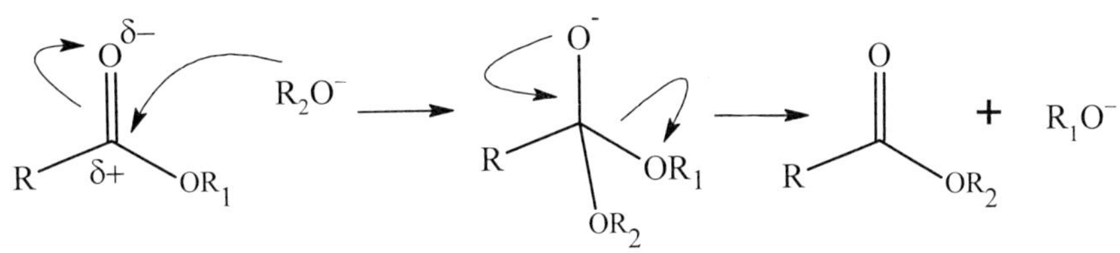

It is very common for a chemical transformation based on acyl nucleophilic substitution, to also involve other reaction steps, most often proton transfers. Thus, conversion of anhydrides to amides is acyl nucleophilic substitution with proton transfers after each step.

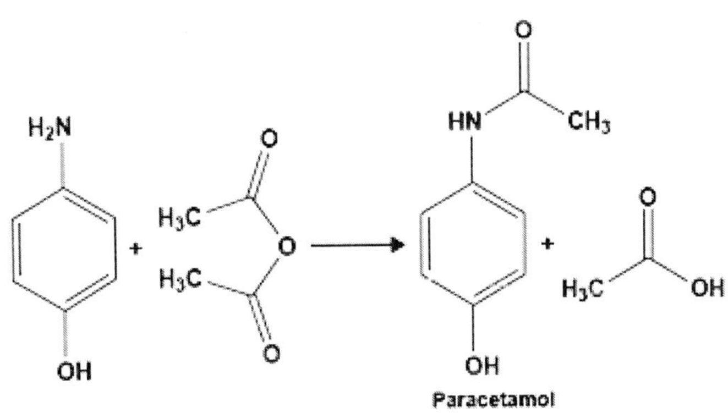

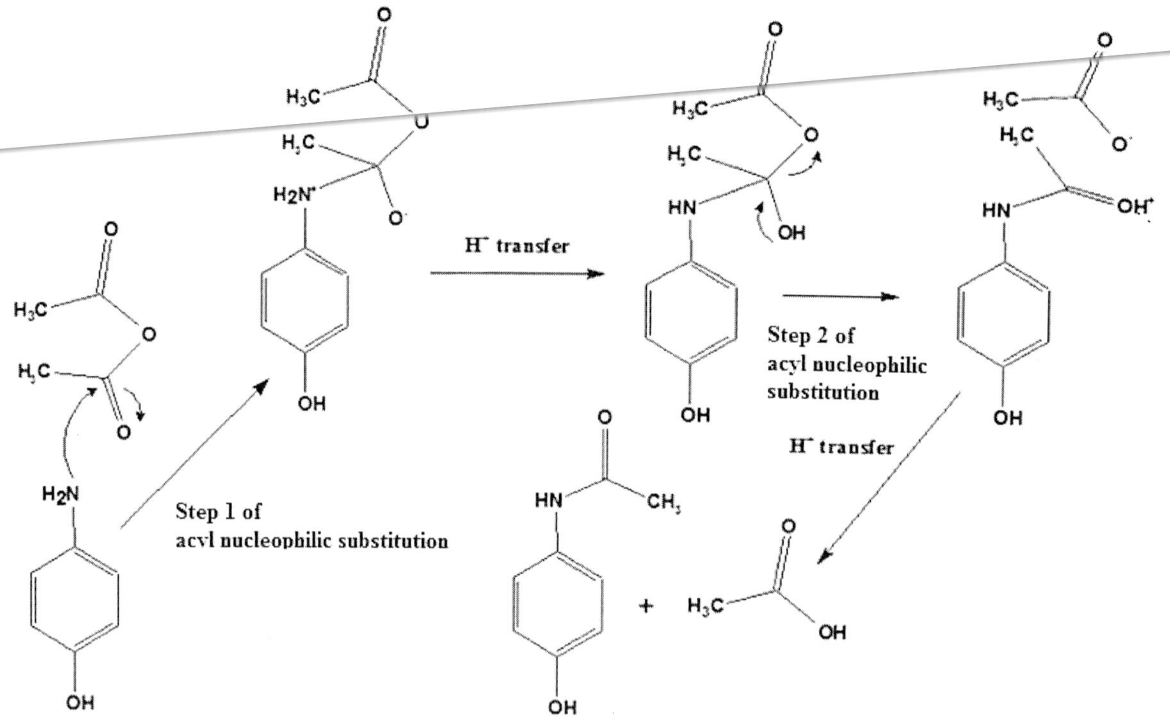

In any full organic chemistry textbook, you will find a plethora of seemingly different reactions of carboxylic acid derivatives. Rather than trying to memorize all of them, first, find the two steps of acyl nucleophilic substitution. All remaining steps will be most likely either proton transfers, or other reactions that you already know. If the nucleophile is a weak base like water, protonation of the carbonyl group activates it toward a nucleophile, which will be the step preceding acyl nucleophilic substitution.

When water or its conjugate base (OH^-) acts as a nucleophile, the nucleophilic acyl substitution produces a carboxylic acid or its salt and thus belongs to the class of hydrolysis reactions. Hydrolysis means "to break with water."

Alkaline hydrolysis of esters (http://ochem.orgfree.com/hydro.html) is also referred to as saponification, is useful in converting fats to soaps.

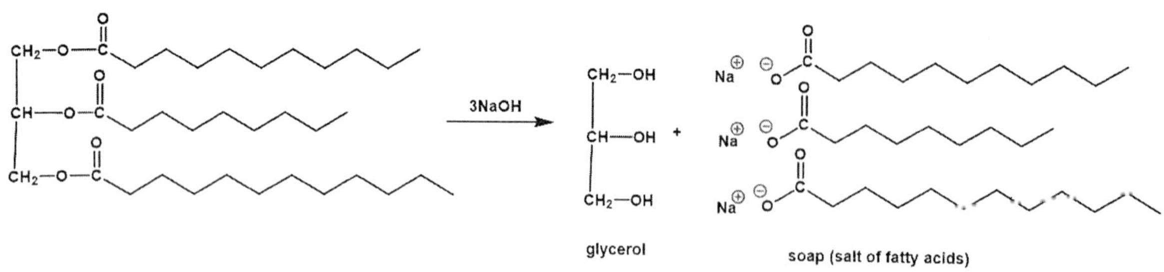

This mechanism is acyl nucleophilic substitution followed by proton transfer.

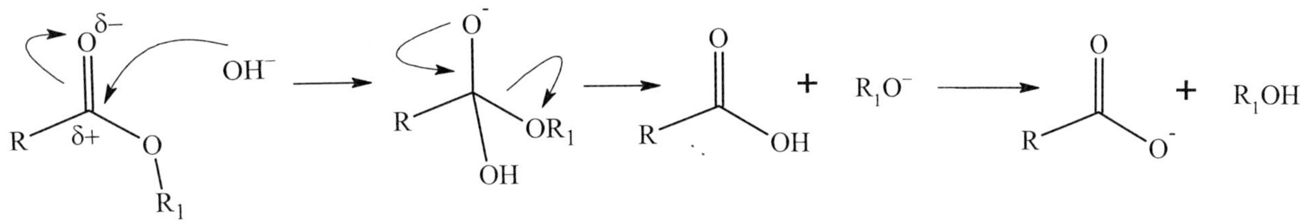

Carboxylic acids, due to their pK_a values, are almost completely ionized at pH = 7. Therefore, carboxylic acids provide most of negatively charged groups to biological systems.

At room temperature, a carboxylic acid protonates an amine, producing an ammonium salt, which provide most of positively charged groups to biological systems.

Heating a mixture of a carboxylic acid and an amine at high temperatures, or their reaction at room temperature in the presence of coupling reagents as most common industrial approach to the synthesis of amides.

Carboxylic acid chlorides are prepared by the reaction of carboxylic acids with thionyl chloride, or PCl_5, or other chlorides of inorganic acids.

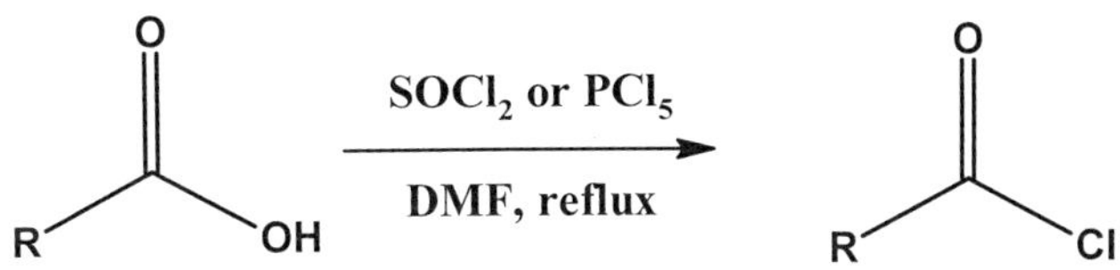

Therefore, any carboxylic acid derivative can be converted to any other carboxylic acid derivative by hydrolysis to the carboxylic acid, followed by its conversion to the corresponding acid chloride, which finally produces the desired product by acyl nucleophilic substitution.

Fischer esterification (http://ochem.orgfree.com/Ester.html) is the reaction between a carboxylic acid and an alcohol in the presence of a strong acid catalyst, which is the most common method to produce esters.

In the previous chapters we have shown that aldehydes and ketones are readily reduced to alcohols by sodium borohydride ($NaBH_4$). Less reactive esters normally do not react with $NaBH_4$. It takes a more reactive reagent, $LiAlH_4$, to reduce esters.

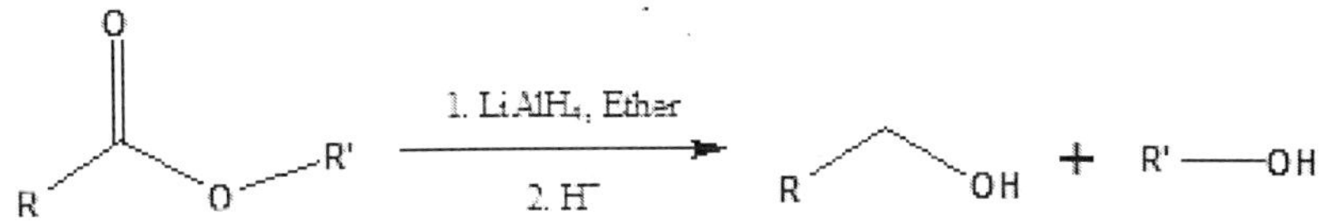

$LiAlH_4$ is also able to reduce any other acyl-containing carboxylic acid derivative.

When a Grignard reagent (a carbon-centered nucleophile) enters the acyl nucleophilic substitution, the reaction results in a ketone, which is more reactive toward nucleophiles than the original carboxylic acid derivative. Therefore, ketones produced by the reaction of a carboxylic acid derivative with the first equivalent of a nucleophile will immediately react with the second equivalent of a Grignard or organolithium reagent.

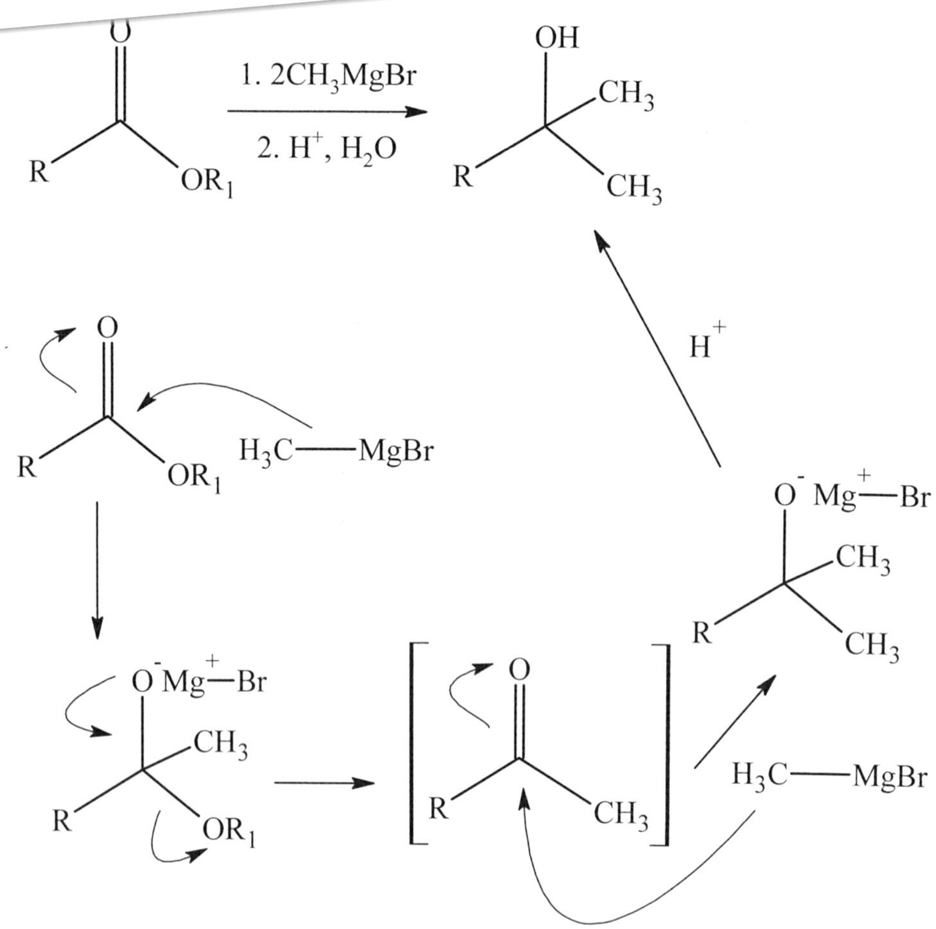

Under certain conditions, a carboxylic acid can lose its carboxy-group, which is naturally termed *decarboxylation.* Any carboxylic acid undergoes decarboxylation when melted with an alkali.

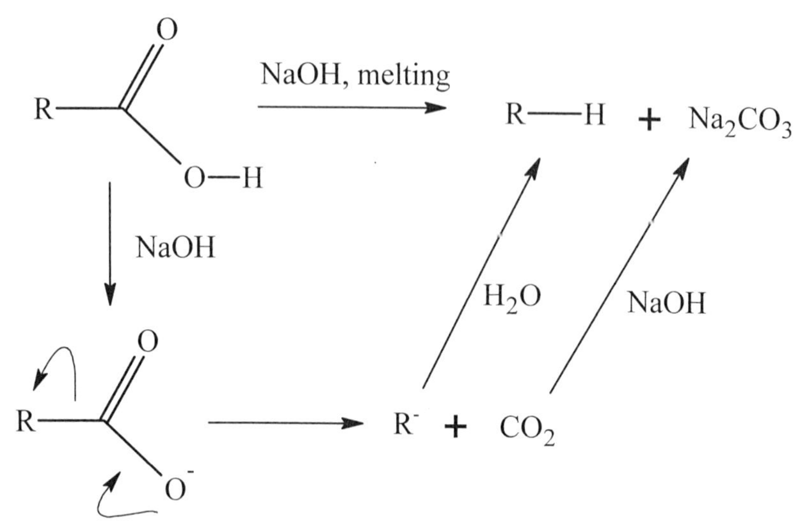

When there is a polar double bond at the β-position to the carboxy-group, decarboxylation proceeds under much milder condition – under 100 °C – because of the intermediacy of a stable six-membered transition state.

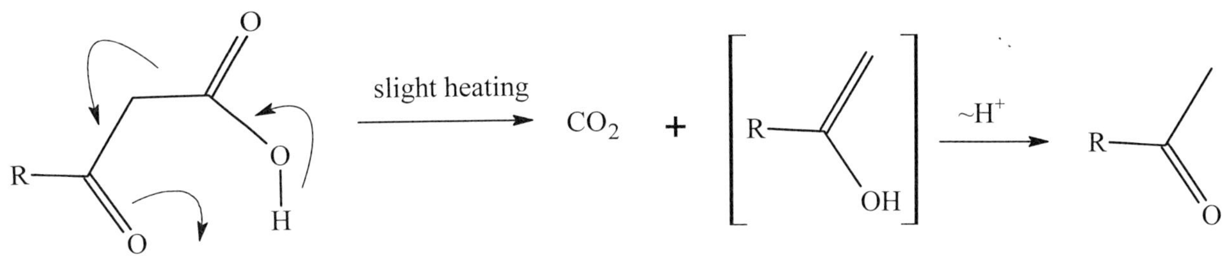

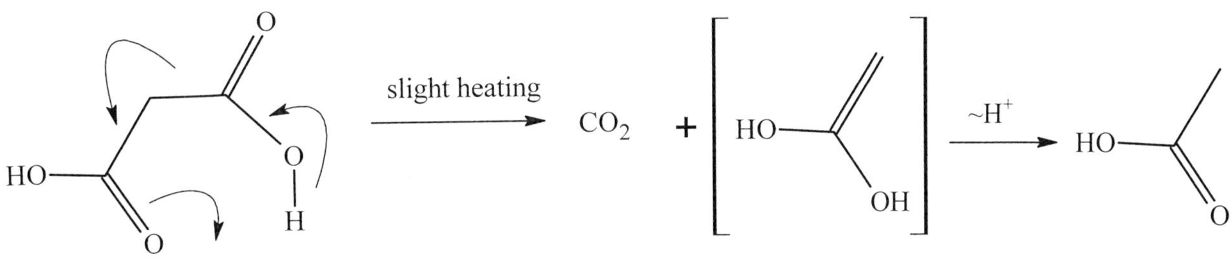

a) What can go wrong?

- The common shorthand notation of carboxylic acids (RCOOH) and carboxylic acid derivatives (for example, esters, $RCOOR_1$) may cause confusion on how exactly the atoms are bonded. This notation should be used only after the students got used to the detailed structure of the carboxy-group.

- Decarboxylation of beta-ketocarboxylic acids is understood better if the mechanism is topologically related to the Diels-Alder addition (http://ochem.orgfree.com/diels.html). Both mechanisms consist of the circular displacement of three electron pairs.

- Reactions with the Grignard reagent are easier to understand, if the C-Mg bond is presented in the ionic form, with the negative charge on carbon, and the O-Mg bond in the tetrahedral intermediate is presented with the negative charge on oxygen.

- The variety of reactions of carboxylic acid derivatives can be overwhelming. The best way to learn each reaction is to find the two steps of acyl nucleophilic substitution. Then, understand the role of other steps in the whole mechanism.

b) Pushing the limits

Comparing basicity of the leaving groups gives us an easy and convenient shortcut to the explanation of the relative reactivity of carboxylic acid derivatives, which luckily works in most cases. However, the reactivity of a carboxylic acid derivative also depends on the interactions in the original carboxylic acid derivative, and in the tetrahedral intermediate. A great resource to take into account those interaction is the Natural Bonding Orbital theory, which provides a much deeper insight into the issue.

c) Practice problems

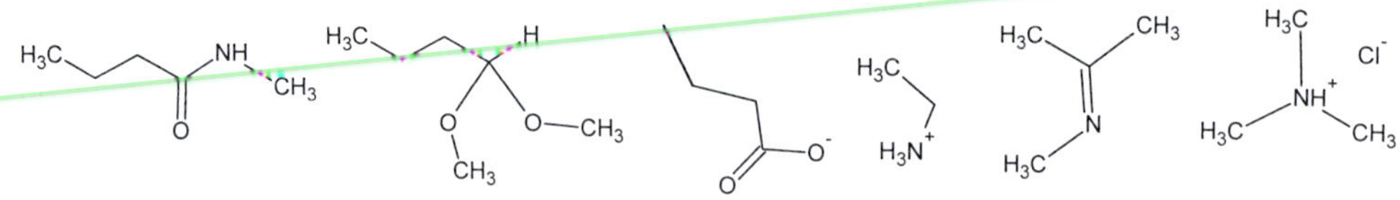

1. **In the compound produced by a reaction between a carboxylic acid and an amine without a catalyst and without heating, circle the methyl group originated from the amine**

2. **Which reagents will produce the compound you selected for question 1?**

   A. Butanoic acid and aminoethane B. Butanal and methanol

   C. Propanoic acid and aminoethane D. Propanone and aminomethane

   E. Trimethylamine and HCl

3. **In the compound produced by a reaction between a carboxylic acid and an amine upon heating or in the presence of a coupling reagent, circle the new covalent bond**

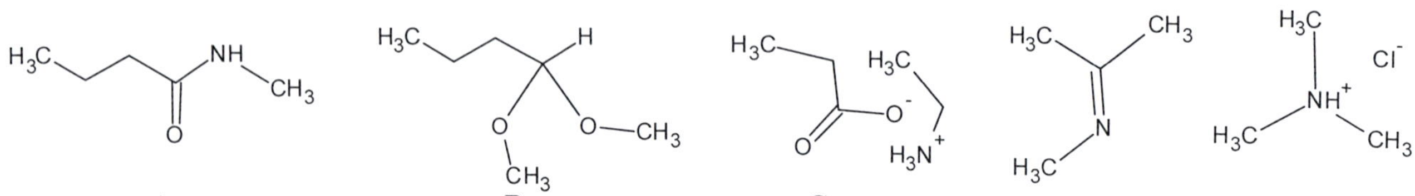

A B C D E

4. **Which reagents will produce the compound you selected for question 3?**

   A. Butanoic acid and aminomethane B. Butanal and methanol

   C. Propanoic acid and aminoethane D. Propanone and aminomethane

   E. Trimethylamine and HCl

5. **In the compound that will produce ethanoic acid upon acidic hydrolysis, click on the N-C bond that remains intact during hydrolysis**

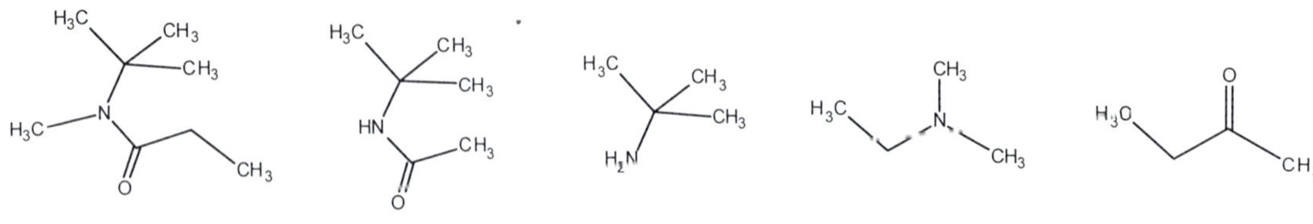

A B C D E

6. **Which reagent converts acetic acid to a reagent capable of acylating toluene by the electrophilic substitution?**

   A. Methylamine B. Methylamine / DCC C. PCC D. $SOCl_2$ E. HCl

7. **Phenyl magnesium bromide reacts with both acetone, and ethylacetate. Which step of the mechanism is taking place in both reactions?**

A. The first equivalent of phenyl magnesium bromide adds to the carbonyl group

B. The tetrahedral intermediate expels the ethoxy-group and restores the carbonyl group

C. The second equivalent of phenyl magnesium bromide adds to the carbonyl group

D. The tetrahedral intermediate expels the methyl-group and restores the carbonyl group

E. Protonation of the tetrahedral intermediate produces the final tertiary alcohol

8. **Which reagent sequence of reagents will convert acetamide to ethyl acetate?**

A. C_2H_5OH, OH^- B. C_2H_5OH, H^+ C. 1. H_2O, H^+; 2. $SOCl_2$

D. 1. H_2O, H^+; 2. C_2H_5OH, H^+ E. 1. $NaNH_2$; 2. C_2H_5OH, H+

9. **The reaction of phenylmagnesium bromide with methyl benzoate followed by acidification produces triphenylcarbinol. What are the sources of phenyl groups in triphenylcarbinol?**

A. One phenyl group from methyl benzoate and two phenyl groups from phenylmagnesium bromide

B. Two phenyl groups from methyl benzoate and one phenyl group from methyl magnesium bromide

C. All three phenyl groups from phenylmagnesium bromide

D. All three phenyl groups from methyl benzoate

10. **What is the product of reduction of ethyl 4-oxobutanoate with sodium borohydride in ethanol at room temperature for 30 min?**

A. Ethyl 4-hydroxybutanoate B. 1,4-Butanediol C. 4-Hydroxybutanal D. Ethyl butanoate

d) Making connections

This chapter introduces one more type of nucleophilic substitution in addition to those you leant so far (S_N1, S_N2, S_EAr, S_NAr, benzyne mechanism). The first step of the acyl nucleophilic substitution is nucleophilic addition that you learnt for aldehydes and ketones. Understanding of the mechanism of decarboxylation of β-ketocarboxylic acids is greatly facilitated by the prior knowledge of the concept of enolization and the mechanism of Diels-Alder addition. Please, note, that the next chapter should not be started without a good grasp of the material presented here.

Chapter 10

Reactions of enols and enolates.

From the two previous chapters, you learned the role of the carbonyl group in the reactions of aldehydes, ketones, and carboxylic acid derivatives with nucleophiles. Here we will focus on the ability of carbonyl compounds themselves to form nucleophilic particles – enoles and enolates. The carbon atom directly bonded to a carbonyl group is called an α-carbon and the hydrogens bonded to an α-carbon are referred to as α-hydrogens (*alpha*-hydrogens). Compounds with alpha-hydrogens have elevated acidity, because their conjugate base is stabilized due to the conjugation with the C=O group.

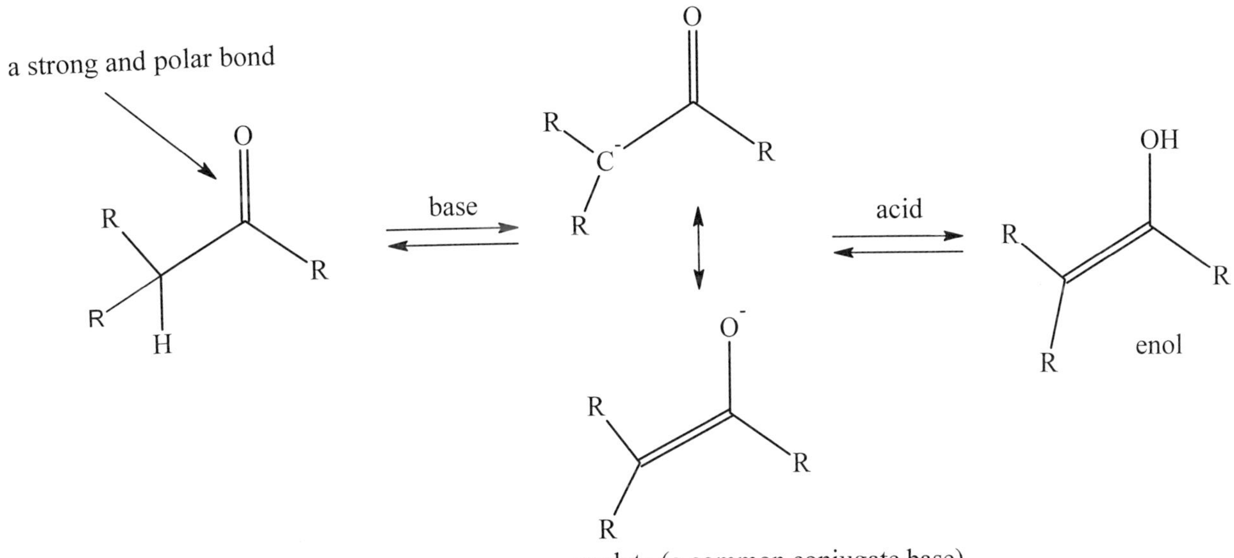

enolate (a common conjugate base)

This conjugate base is termed *enolate,* whose protonation at oxygen produces an *enol.*

As a Brønsted base, the enolate can be protonated either at the oxygen or carbon, giving rise to either *enol* or keto form, respectively. Therefore, an enol and its keto form (tautomers) are sharing a conjugate base (enolate), and usually are rapidly interconverting in the presence of a trace amount of either base, or acid. This equilibrium is an example of *tautomerism – a rapid interconversion between constitutional isomers.*

Most aldehydes and ketones exist predominantly in the keto form. Propanone, for example, is 99.9997% in the keto form and only 0.0003% in the enol form. The main reason for the greater stability of the keto form is the higher bond energy of the C=O and C-H bonds in the keto form compared to the weaker C=C and O-H bond energies in the enol form.

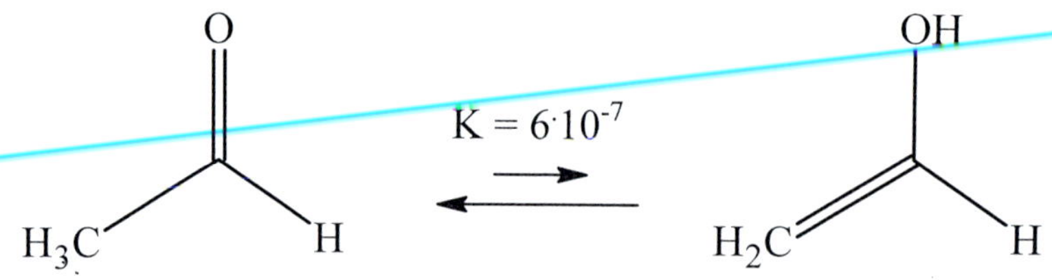

If there are two carbonyl groups separated by a single CH_2 group, the α-hydrogen between the two carbonyl atoms will become even more acidic because of the resonance delocalization of charge onto the two carbonyl oxygens. This type of compounds is termed "β-dicarbonyl compounds", because the second carbonyl group belongs to the carbon second next to the first carbonyl group. The enols of β-dicarbonyl compounds are additionally stabilized by an intramolecular hydrogen bond.

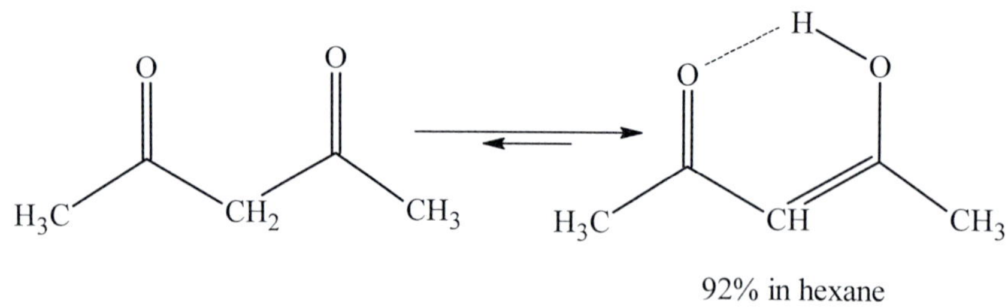

Because of the resonance donor effect of –OH and –O⁻ next to the C=C bond, enols, and especially, enolates, are very reactive with electrophiles. As the enol or enolate concentration depletes, the equilibrium shifts, converting more and more of the carbonyl tautomer to the product. Thus, enols and enolates account for a significant portion of the reactions of carbonyl compounds.

For example, the enol form of a carbonyl compound rapidly reacts with molecular bromine (Br_2). In the next step, this intermediate deprotonates to the final α-brominated carbonyl compound.

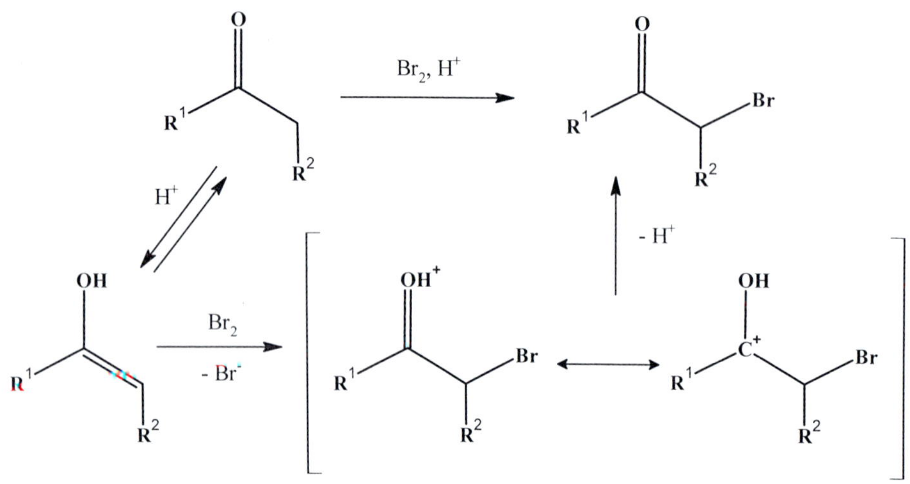

Acid-catalyzed enolization

Resonance-stabilized intermediate

Intermediacy of the enol limits bromination to the α-positions with respect to the carbonyl group, because the C=C bond in the enol cannot extend by this mechanism further than the α-carbon.

The steric hindrance introduced to the molecule by the atom of bromine substantially slows down further bromination, so the monobrominated product usually can be isolated in decent yields. Chlorination proceeds similarly to bromination.

In the presence of a base, the reaction course significantly changes. First, highly nucleophilic enolate rather than enol undergoes bromination. Thus, after the first bromination, high nucleophilicity of enolate overrides the steric hindrance, so all the remaining α-hydrogens are rapidly substituted with bromines.

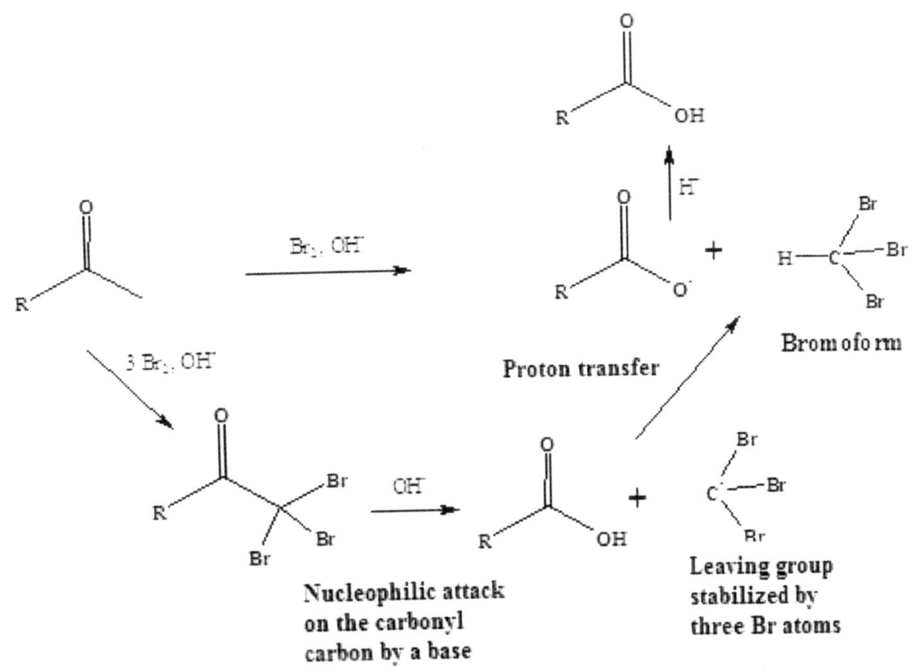

Halogenation of aldehydes and ketones containing the acetyl group (CH_3CO) under basic conditions proceeds even further. Substitution of all three hydrogens by bromine produces a good leaving group ($^-CBr_3$), where the negative charge is stabilized by three electronegative halogens. As a consequence, the final product of bromination easily undergoes nucleophilic acyl substitution, just like a carboxylic acid derivative, producing the typical products of hydrolysis under alkaline conditions: a carboxylate and a haloform (CHX_3).

The haloform reaction proceeds with chlorine and iodine as efficiently as with bromine. Since iodoform (CHI_3) is solid, iodoform reaction (I_2, OH^-, H_2O) is a good quantitative test for methylketones.

Carboxylic acid also undergo halogenation to the α-position (Hell-Volhard-Zelinsky (HVZ) reaction). However, this reaction requires presence of a catalyst, most commonly, $PHal_3$.

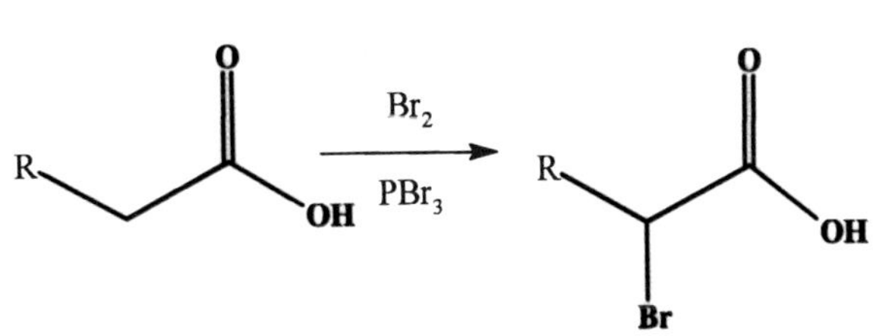

The Aldol Reactions of Ketones and Aldehydes

Aldol addition is addition of the α-CH-bond of an aldehyde or ketone to the carbonyl group of another molecule of an aldehyde or ketone. Dehydration of the product of aldol addition to an α,β-unsaturated carbonyl compound completes the sequence of the *aldol condensation*.

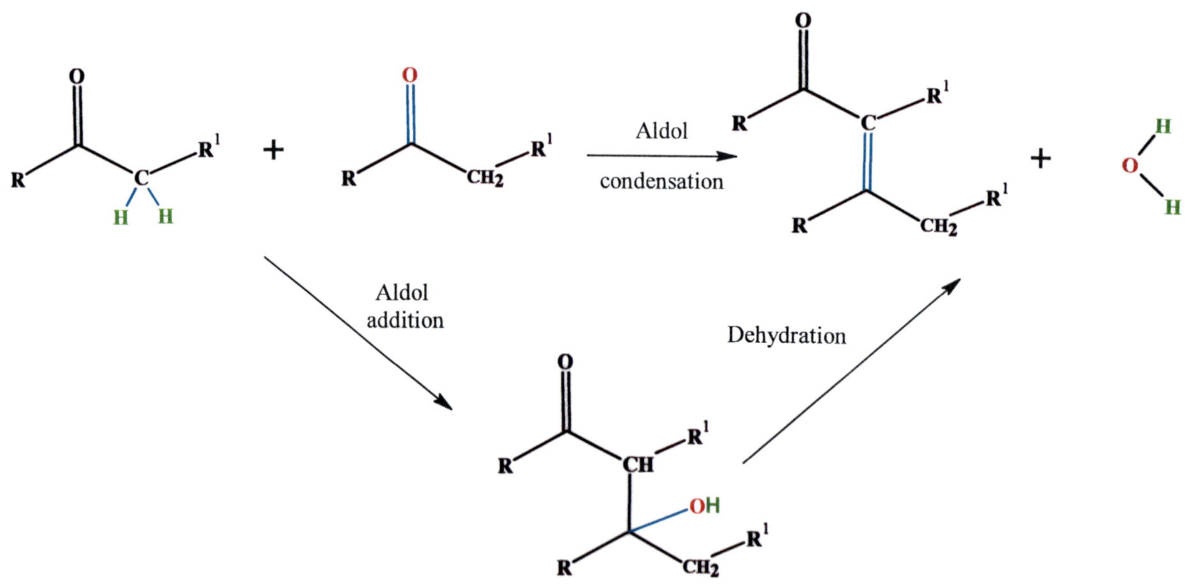

This figure suggests a useful shortcut to derive structure of the product of aldol condensation: 1) Remove oxygen from the carbonyl group of the source of electrophile; 2) remove two hydrogens from the α-position of the source of nucleophile; 3) Connect two pieces with a C=C bond.

Both aldol addition and aldol condensation can be catalyzed by either base or acid. The aldol addition is more likely to continue as condensation under harsher conditions (elevated temperature, higher concentration of an acid or base, or additional stabilization of the C=C bond in the product). Mechanistically, the aldol reaction is driven by the interaction of a nucleophile (enol or enolate) and an electrophile (the carbonyl group).

Claisen condensation is acyl nucleophilic substitution in an ester, where the nucleophile is an enolate generated from the same ester.

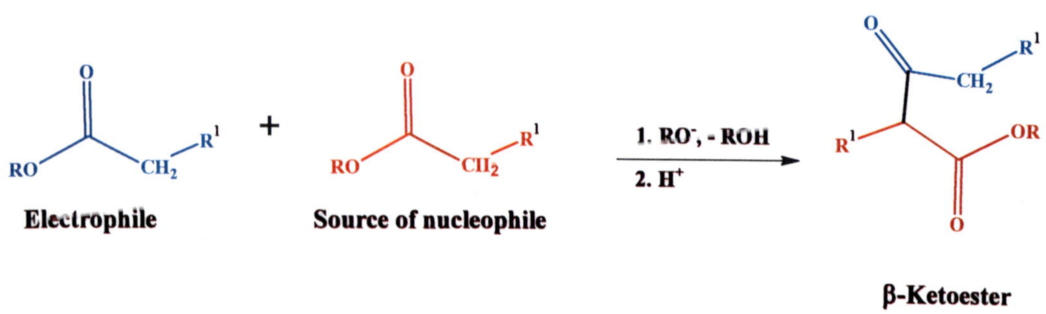

Similarly to the aldol reaction, Claisen condensation leads to the formation of a new C-C bond when an electrophile and a nucleophile originating from carbonyl compounds, interact. However, Claisen condensation does not take place under acidic conditions. Further, esters undergo nucleophilic acyl substitution with bases, which significantly limits our choice of the base-catalyst. The most commonly used base for Claisen condensation is the alkoxide of the alcohol forming the ester. The mechanism of

Claisen condensation is shown in the following scheme.

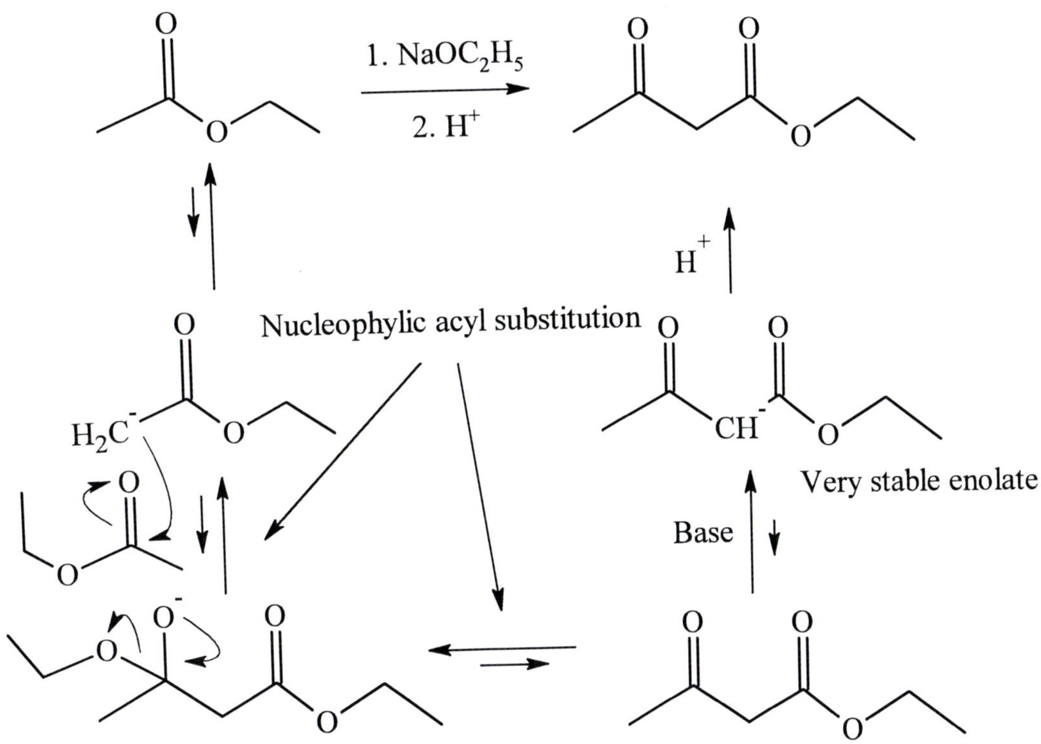

A modification of Claisen condensation (with thioesters instead of esters and an enzymatic leaving group –S-CoA instead of alkoxide and a base) is a very common step in biosynthesis that adds two more carbons to the carbon chain.

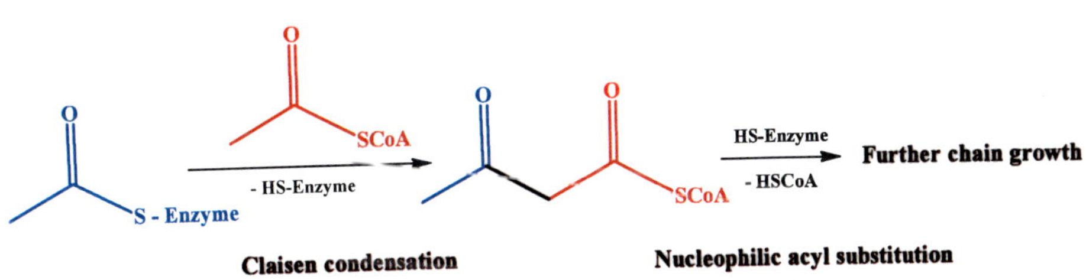

Claisen condensation produces a compound with a CH- or CH_2-group between two carbonyl groups. This type of compounds is termed *β-ketoesters*, which has two practically useful features. First, the hydrogen attached to the carbon between two carbonyl groups has enhanced acidity, which significantly facilitates formation of a nucleophilic enolate. Second, hydrolysis of the ester group produces a β-carbonyl carboxylic acid, which easily undergoes decarboxylation and forms a ketone. Both hydrogen atoms of the CH_2-group can be sequentially substituted with different alkyl groups, making *β-ketoesters* valuable precursors of ketones.

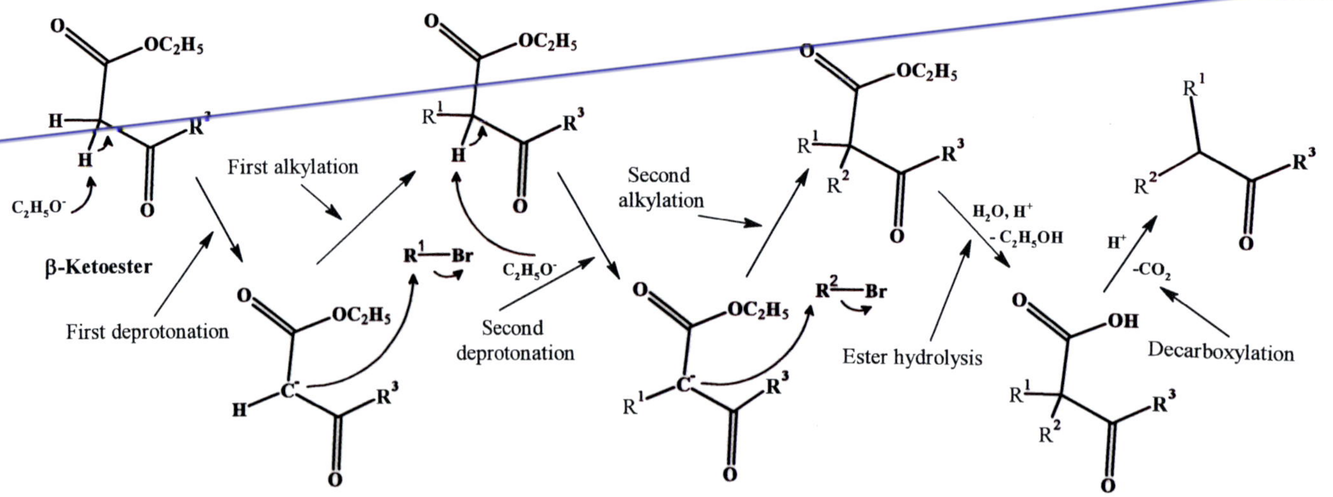

What can go wrong?

Attempting to memorize all these reactions without understanding the common mechanistic steps is rarely productive. Please, remember that all reactions of enols or enolates include the key step (interaction of a nucleophile with an electrophile) and preparatory steps (protonations or deprotonations at the beginning, middle, or/and at the end of the reaction sequence). Under basic conditions, enolates act as nucleophiles, and carbonyl groups act as electrophiles. Under acidic conditions, enols act as nucleophiles, and protonated carbonyl groups act as electrophiles.

Pushing the limits

The carbonyl groups are able to react with a wide variety of other nucleophiles: "enolates" not derived from a β-ketoester or β-carboxyester (nitro-, sulfoxy-, phosphonato-derivatives), acetylides, Grignard reagents. In the aldol and Claisen-type reactions, the nucleophile and electrophile can be generated from different compounds, expanding the-type variety of products. Placing the newly formed carbon-carbon bond correctly between the nucleophilic and electrophilic reactants is a key to understanding those transformations.

Practice problems

1. **Which pair of reactants will undergo aldol condensation to produce methylvinyl ketone?**

   A. Propanone and ethanal **B.** Methanal and propanone
   C. Two molecules of propanone **D.** Methanal and ethanal
   E. Two molecules of propanal

2. **Which of the following compounds cannot be a source of a nucleophile in the aldol addition?**

   A. Propanone B. Ethanal C. Methanal D. 2-Butanone E. Butanal

3. **Condensation of which starting materials in the presence of KOH will lead to 3-phenylprop-2-enal** (also known as cinnamaldehyde that gives cinnamon its flavor)?

   A. Ethanal and methanal **B.** Ethanal and benzaldehyde
   C. Methanal and benzaldehyde **D.** Propanal and benzaldehyde
   E. Ethanal and propanal

4. **Which of the following compounds will give a negative iodoform reaction?**

   A. Propanone **B.** 2-Butanone **C.** Ethanal **D.** 3-Pentanone **E.** 2-Pentanone

5. **Which of the following represents a step of Claisen condensation?**

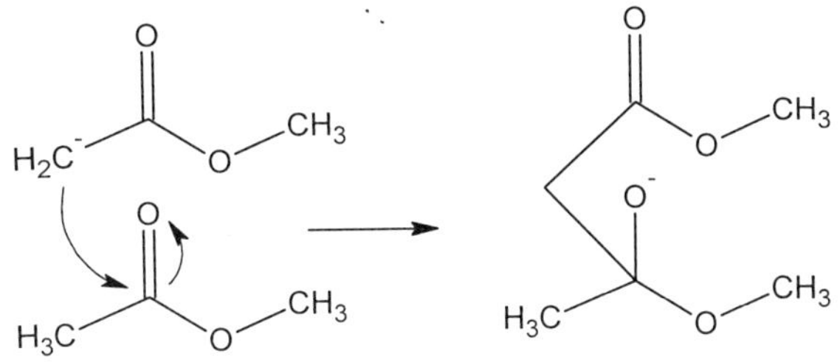

A

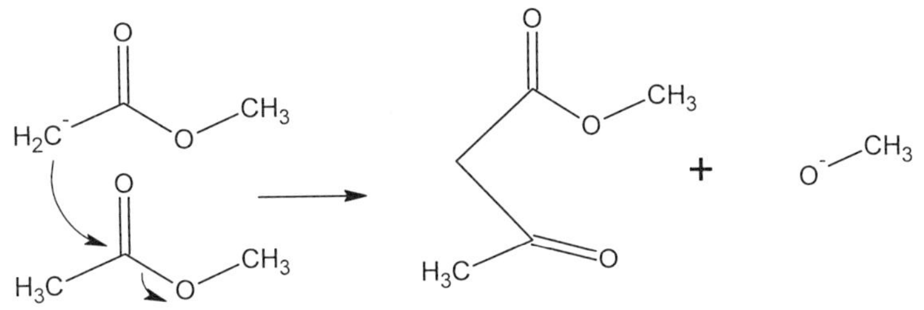

B

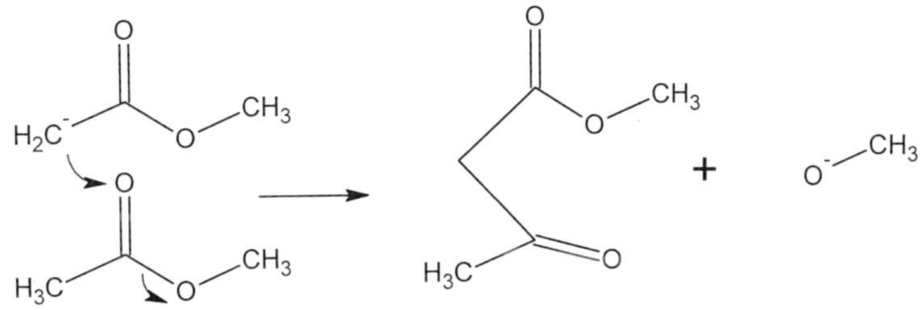

C

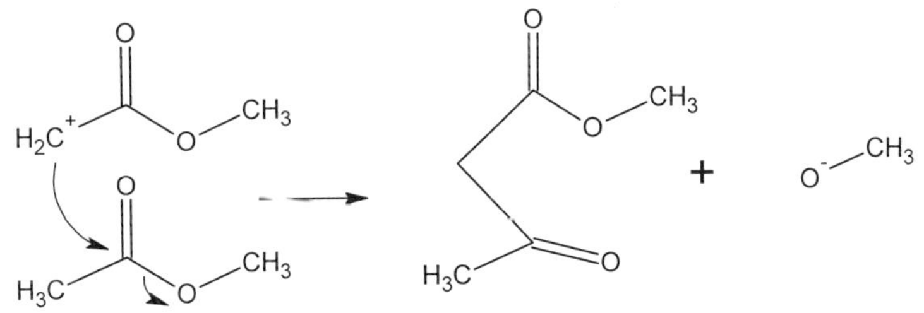

D

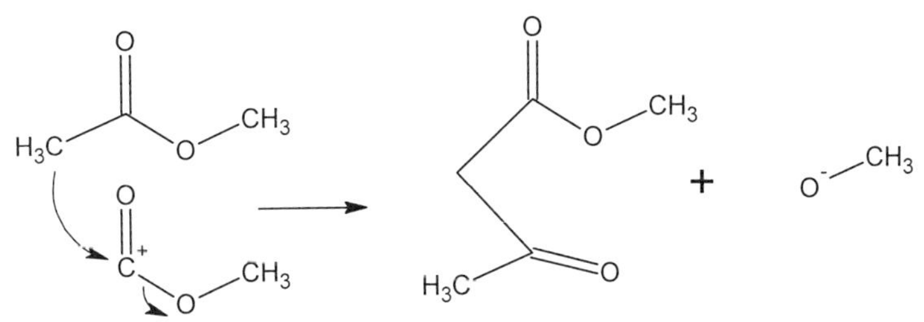

E

6. **Which step of the following reaction sequence includes decarboxylation?**

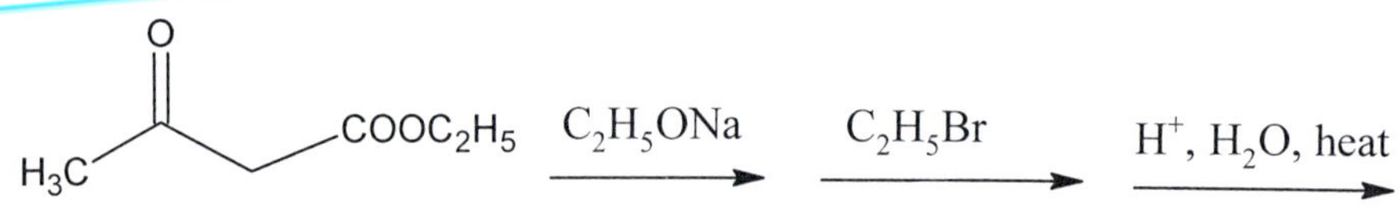

A. Deprotonation **B.** Alkylation **C.** All steps

D. Heating with acid **E.** Neither step

7. **Which of the following compounds <u>is not</u> the major compounds of the keto-enol equilibrated mixture?**

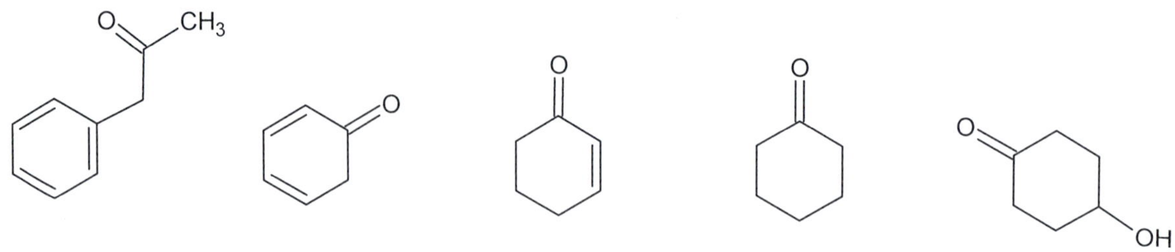

A **B** **C** **D** **E**

8. **Which step of the following reaction sequence leads to formation of a new C-C bond?**

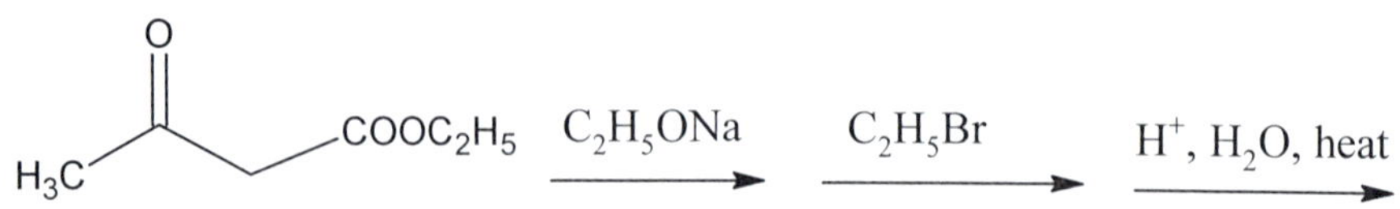

A. Deprotonation **B.** Alkylation **C.** Hydrolysis

D. Decarboxylation **E.** Carboxylation

9. **Which of the following compounds is the major product of the following reaction sequence?** Do not forget reactions of Claisen condensation, hydrolysis of esters, and decarboxylation of β-oxocarboxylic acids.

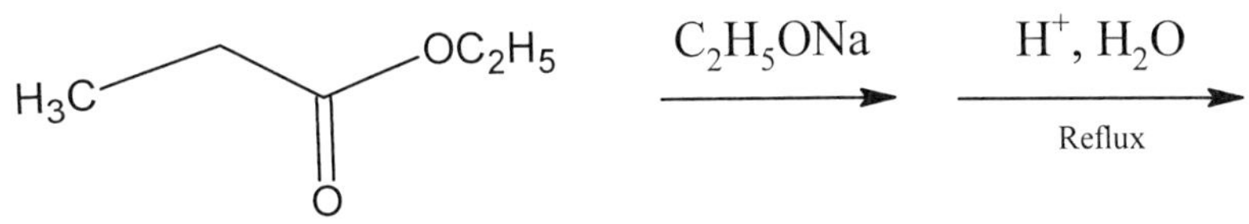

A. 3-Hexanone B. 4-Hexanone C. 2-Pentanone D. 3-Pentanone E. Pentanal

10. **Which of the following compounds is the starting compound of the following reaction sequence?** Do not forget reactions of PCC oxidation and aldol condensation.

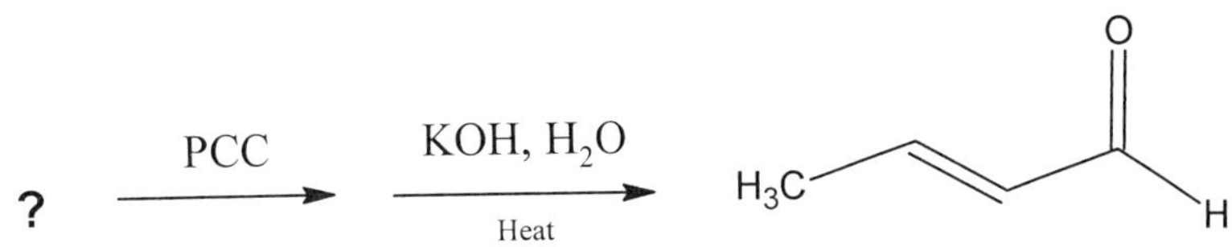

A. Methanol B. Ethanol C. Ethanal D. Pentanol E. Pentanal

Making connections

Studying this chapter lets you creatively apply the knowledge you learnt from the rest of this book. Now you are ready to explore organic chemistry in-depth using full-sized organic chemistry texts. If at some point you get confused with the new material from any source, it is always a good idea to go back to this book, navigate through the *"Making connections"* and *"What can go wrong"* sections to identify and solve the problematic issue.